Chasing the Demon

Chasing the Demon

A SECRET HISTORY OF THE QUEST FOR
THE SOUND BARRIER, AND THE BAND OF
AMERICAN ACES WHO CONQUERED IT

Dan Hampton

WILLIAM MORROW
An Imprint of HarperCollins*Publishers*

HarperCollins books may be purchased for educational, business, or sales promotional use. For information, please email the Special Markets Department at SPsales@harpercollins.com.

A hardcover edition of this book was published in 2018 by William Morrow, an imprint of HarperCollins Publishers.

FIRST WILLIAM MORROW PAPERBACK EDITION PUBLISHED 2019.

Image on title page courtesy of NASA

Photograph by John Carnemolla/Shutterstock

Library of Congress Cataloging-in-Publication Data has been applied for.

ISBN 978-0-06-268873-6

19 20 21 22 23   10 9 8 7 6 5 4 3 2 1

For Colonel Ken "K. O." Chilstrom and
his 114 fallen brothers-in-arms

Contents

Author's Note

For simplicity, I have generally used the U.S. military equivalent when discussing foreign military ranks and command structure. Use of German, Italian, or Japanese words and phrases has been reduced to the bare minimum as they tend to confuse most readers, though certain distinct unit names have been retained. Much of this book concerns the U.S. Army Air Corps and the USAF test program, so I must ask forgiveness from my brothers with gold wings for limiting the numerous contributions of Navy and Marine aviators. There have been scores of superb pilots from both services involved with flight test, but as carrier aviation is essentially tactical in nature, flying supersonic did not obsess naval aviators as it did the Air Force.

I also beg forbearance from Second World War historians for my abbreviated treatment of pivotal operations in the Pacific, Mediterranean, and the savage fighting throughout Europe toward the close of the war. These were history-altering events about which scores of superb books have been written, but I would remind my colleagues that in this work the war is the

historical backdrop that created the men and fostered the aviation advances that make this story possible.

Yet we must not forget that progression into the supersonic
age of flight was undeniably facilitated by World War II, and
the men who chased this particular demon were products of
that war and the unsettled decades preceding that conflict. So
to know them, and to understand the technology and challenges they faced in attempting to fly past the speed of sound,
it is necessary to understand something of their world and its
consequences, both personal and public. In bringing this to
life I have had an incalculable asset: Colonel Ken Chilstrom.
World War II fighter pilot and veteran of the North Africa,
Sicily, and Italian campaigns; graduate of the initial U.S.
test pilot course; and chief of the Wright-Patterson Fighter
Test Division when the XP-86 and Bell X-1 were being put
through their paces. As an author, having a firsthand, eyewitness source to the times and events that shaped our world is
phenomenal. As a fighter pilot myself, to sit and listen to an
aviation legend who has truly "been there and done that" was
awe inspiring. To call him my friend is an honor I will always
cherish.

Keep in mind also that just as the birth of modern flight
is certainly not a single event, neither is the advent of supersonic flight. True, it was the Wrights at Kitty Hawk on that
cold December morning in 1903 who first achieved powered,
controlled flight in a heavier-than-air machine, but note the
qualifiers. Many other men had flown in one fashion or another, and for centuries they had used gliders, with various
types of rudimentary control, to achieve a type of flight. Still
others were able to power their craft off the ground but had no

control whatsoever over the craft once it was airborne. Later, men attempted to blend what they had learned—to use self-produced power—and it was at this point, along with positive control, that true flight was finally achieved. It is precisely the accomplishment of controlled, manned flight under power that justifiably gave the Wrights claim to the title. Other titles and popular claims to fame may be less certain and we shall take an objective look into this.

Throughout the book I have endeavored to distill aerodynamic concepts into a digestible form for those readers without aviation or engineering backgrounds, and I have included the historical aspects of man's quest to fly past the speed of sound for those without historical inclinations. My hope is that there is a story here for all of us who share a fascination with aviation.

As for the demon . . .

It existed, and still does. It lives out just beyond the thin air, elusive and tempting, drawing us further and deeper into his domain. A world of unknowns and lofty, dangerous pursuits such as high-altitude flight, global circumnavigation, and, yes, the so-called sound barrier. The latter term is a romantic label for flight beyond the speed of sound, and as such it is attractive to moviemakers or publishers. It has a ring to it, an air of satisfying finality, nearly an absolute, which can only be overcome through extreme courage and skill. There is truth in this, yet the term is utterly artificial for aviators, engineers, or historians, at least in the intellectual sense: no such barrier actually existed.

Or did it?

From another viewpoint the barrier and its demon lived in men's imaginations and would continue to do so until both man and machine evolved sufficiently to dispel the myth. Indeed, there *was* a physical barrier, but it was of man's own making through his misunderstanding of transonic aerodynamics, and the propulsion systems required to push him faster. Once these were achieved the actual event of muscling through Mach 1 was anticlimactic: an inevitable moment in time. But it is here, in the shaping of the men and the struggle to develop the science, that the real story lies. The story of those who chased the demon, in any of his forms, far out into the unknown only to see his shape fade elusively into another undiscovered realm of his world. The demon beckoned men to follow, and still does today, to continue chasing him toward whatever else is out there.

Why were men striving to fly past the speed of sound, past a known, calculated point that, for many, had become a barrier to further flight? How did such a time in history arise that would permit such an endeavor? What caused the very fertile ground of the 1930s and 1940s to be so fertile, and how did the global stage come to be *the* stage upon which this drama was set? Simply phrased—why then? And equally important, how did the men who chased the demon become the men they were? What events shaped their inherent talents and abilities to the point where the demon of speed—and others—could be successfully pursued?

Lots of questions.

To answer them we must look well beyond the very short, four-minute rocket burst that *officially* took man beyond the speed of sound during October 1947, into what made that flight, and very likely others before it, possible. We must look at war and peace, politics, science and technology, to examine factors that molded the world we inherited and that still impact our lives today. As with other seminal developments we take for granted today, the origins and motivations have often become obscured, idealized, or, worse still, forgotten altogether. At several points in the history of flight and its subsequent quests, credit has not been given to those it is due, but rather to those with the best publicity. I certainly do not impugn those who have gone before; they were brave men who purposely took extraordinary risks, albeit for reasons as different as the pilots themselves.

This work will simply lay out various facts, unburnished by time and without the glitter of legend, to at least encourage readers to pause for thought. To reflect that just because we were taught a thing it does not necessarily mean it is true, and perhaps to remind ourselves that much of what is accepted as historical fact is, in fact, neither history nor fact. Such a luxury is only possible because others have gone before us and done the dirty work, so to speak—men who offered their reputations, suffered the slings and arrows of their peers, and often lost their lives in the pursuit of the unknown.

The unknown; it is an ideal to some, a demon of sorts to others, and always a challenge. There were those who believed the demon in the thin air past the speed of sound did not exist at all, but *something*, most admitted, was out there. Something that locked up controls, shook aircraft apart, and killed men. Was it only the demon of speed or were there others, and are there still more out there waiting to be found? In chasing this particular demon I have discovered that there is not a single entity; or maybe there is, but it is one who can adopt many faces. A demon of altitude; of power; of war; and, certainly, one of speed. A demon who retreats as quickly as man, with his boundless optimism and eternal cleverness, advances.

—Dan Hampton
Hanover, New Hampshire
March 2018

Chasing the Demon

Prologue

I t looked like a crucified man.

Stark and lonely, the little aircraft hung in the chill morning air of March 1, 1945, pointed skyward on a seventy-nine-foot latticed launcher. Its stubby wings, nearly twelve feet across, gave a peculiar, truncated appearance compared to the graceful lines of its propeller-driven cousins. Even here in the Heuberg Military Training Area, where strange sights were normal, it was an odd craft.

Called a *Natter*, it did, in fact, resemble the viper from which it took its name, down to the two dozen nose-mounted Hs-217 missiles that gave it a lethal bite. Arranged in a *bienenwabe*, or honeycomb, even one of the 73 mm projectiles could bring down an American heavy bomber, and this was the Natter's sole purpose. By 1945 German air defense fighter units were losing almost as many aircraft per month as they were supplied and far, far too many pilots who could not be replaced. It was an unsustainable situation.

It meant defeat.

The Natter, and weapon systems like it, were intended to change that. Extensive wind tunnel testing had been concluded in September 1944, with a dozen launch and glide tests successfully completed over the next five months. Excellent stability was recorded, and the controls were "light and well coordinated" with no sideslip or yawing.* According to Hans Zuebert, the program's senior test pilot, the "handling and flying qualities were superior to those of any of the standard German single-seat fighters." So when Lothar Sieber, the Natter's twenty-two-year-old Luftwaffe test pilot, clambered into the tight, spartan cockpit, he was confident of being the first human to fly a guided rocket.

But it was not a suicide weapon.

Unlike their Japanese allies, Germans were not inclined to perish willfully, and they knew that while each Natter could be constructed in a matter of days, it took twenty years to raise and train a flyer. Each man, like Lothar Sieber, was a volunteer and a trained fighter pilot who intended to fight another day. Given his value, considerable thought went into sparing the aviator's life. Solid armor plate lined the forward and rear bulkheads, while sandwich armor protected him from the sides.

Technically known as the Ba-349, the Viper was made from wood and fastened with nails or glue. Scarce metal was sparingly utilized for the load-supporting attachment joints, control push-rods, fuel lines, and the motor. Its construction was simple—even crude—by the standards of the day and would

---

* A glossary of potentially unfamiliar aviation and military terms appears on page 271.

never withstand prolonged use. Indeed, each would not survive its maiden flight nor was it designed to do so. These days were dark for Germany, and standards had fallen dramatically. This aircraft had to be made with a minimum of low-grade material in poorly equipped shops by inexperienced, amateur workers and it certainly was. Wonder weapons, like the Natter, were a desperate last gamble intended to slow the Allied advance and at least open negotiations for peace.

Always pressed for resources, the Reich was in perilous shape by 1945. Barely 1,000 man-hours were required to construct a wooden Natter with only basic tools compared to roughly 4,000 man-hours necessary to produce a Me 109G in 1944. If a single Natter could bring down at least one Allied B-17 (which required 18,600 man-hours to build) and put its crew of ten trained men out of the war, then the economics made sense to the Germans.

The concept was simple: each aircraft would fly only once, deliver its lethal bite, and then die with several big bombers and their crews. Warned of the altitude, airspeed, and heading of incoming enemy formations, a Natter battery of ten aircraft would be assigned the initial attack. Angle and azimuth information was fed into an anti-aircraft computer, usually a Kommandogerat stereoscopic flight director normally employed with 88 mm FLAK batteries. Utilizing a simple linear speed calculation, the launcher was aligned in azimuth, and the aircraft's elevons were locked at the proper deflection angle computed to complete the intercept.

Based on its astounding 35,800-feet-per-minute climb, as well as the known distance and altitude of a bomber formation, all four Schmidding solid-fuel booster rockets were electronically fired.* Rising ballistically to approximately 500 feet, the Natter now had sufficient airspeed for the control surfaces to function. After ten seconds the boosters were jettisoned, the Walter HWK 109-509 liquid-fuel motor kicked in, and the cockpit controls were unlocked, enabling the pilot to manually fly the rocket. However, flight data was now being continuously radio linked to the Patin three-axis autopilot, and the idea was to let the system complete the intercept until the pilot visually sighted the bombers. At this point he would override the autopilot, blow off the Perspex nose cover protecting the rockets, and close to firing range. Ripple firing two dozen spin-stabilized rockets into a tightly packed formation of American B-17s would have, it was hoped, a catastrophic effect.

Given the aircraft's tiny size, nose on aspect, and astonishing top speed of nearly 600 miles per hour, the Germans believed such an assault was indefensible. Following his attack, the pilot would glide away from the mangled bombers and their scattered escorts. Disengaging the safety on the nose release mechanism, he could jettison the nose section and all the cockpit forward of his seat. As the debris falls away into the high-speed airstream, two half-inch steel cables yank an extremely sturdy parachute from a cavity in the tail section. The body of the aircraft violently decelerates when the chute opens and this flips the pilot forward out of his seat. Once clear, he then activates his personal chute and drifts to safety; the entire flight, from launch to pilot ejection, would last less than five minutes.

In theory.

But the true combat effectiveness of this particular weapon

* Each motor generated 1,102 pounds of thrust.

would thankfully never be known. During that brief flight in 1945, Lothar Sieber did indeed become the first man to be successfully launched in a rocket—for fifty-five seconds. At 1,650 feet, some fifteen seconds into the flight, the canopy inexplicably detached and the Ba-349 flipped onto its back while climbing shallowly to 5,000 feet. Rolling inverted, the Natter then dove straight into the ground about four miles from its launch point, barely missing the little hamlet of Stetten am kalten Markt. A partial left leg, a left arm, and fragments of Sieber's skull were eventually recovered from the fifteen-foot-deep crater.

No one knows exactly what occurred that morning, yet when the canopy detached, it could have struck Sieber. A pilot's conscious or semiconscious reaction would be to grab the stick and pull back, so if this occurred the Natter would have ended up on its back. Entering a cloud deck at that speed and attitude would be disorienting, and Sieber would very likely find himself out of control. If this happened, he would try to bail out, though this would be nearly impossible in a 500 mph dive. The fact that his left leg and arm were recovered lends credence to this notion, however, and he may have been trying to climb out of the tiny cockpit when the Natter impacted.

No matter the cause, with the failure of this single manned flight, and with Allied forces converging on the Germany from all sides, a secret list of top German scientists and engineers was passed to U.S. intelligence. With the war ending, the Americans initiated Operations Lusty and Paperclip to prevent this equipment, these men, and their special knowledge from falling into Soviet hands. Over 1,500 were eventually relocated to the United States, including Major General (Dr.) Walter Robert Dornberger, who would eventually become the vice president of Bell Aircraft.

The Natter project was discontinued, yet there are those who were present as Sieber's Natter shot from the clouds, rocket motor screaming at 600 miles per hour and accelerating, who claim they heard the demon's voice: a boom that would later be recognized as flight past the speed of sound.

Six weeks later a man felt the demon's claws and got close enough to smell its breath. On April 9, 1945, two sleek Me 262 Schwalbe jet fighters lifted off from Lechfeld, just west of Munich, and headed up into thin, clear air. Led by Lieutenant Colonel Heinrich Bär, a Luftwaffe *experten* with over 1,000 combat sorties and 220 confirmed victories, the jets leveled off at 36,000 feet. Today was a training mission for Bär's wingman, Hans Mutke, who was converting to the Me 262 after three years of flying Bf 110 night fighters.* The Schwalbe, or "Swallow," was now Germany's last realistic hope to alter the drastic situation in the skies over the Reich. Powered by a pair of Junkers Jumo engines, each producing 4,000 pounds of thrust, the Messerschmitt could climb at 3,900 feet per minute and sustain 532 miles per hour in level flight at 26,000 feet. The armed variant usually carried four 30 mm cannons, and twenty-four 55 mm rockets capable of hitting a B-17 from a half mile away. It was a game changer and, if not a war win-

---

* Mutke remained a *fahnrich*, an officer cadet, all through the war because he would not join the Nazi Party. Surviving the war, he finished medical school and became a gynecologist.

ner at this late stage in the conflict, there still were those who believed it was still not too late, which was why experienced, twin-engined pilots like Mutke were being rushed through the conversion course.

"He's under attack . . . right now."

Bär's voice was calm as he rolled over and dove his armed Me 262 left toward a P-51 attacking another German fighter far below. Mutke followed in his unarmed jet with no external tank. Though a highly experienced pilot, this was only his third jet sortie and, as one would in a piston-engined fighter, Mutke left the throttles up. Passing about 35,000 feet in a 40-degree dive, the jet began bucking wildly, and the tail began to yaw. Obviously alarmed, he noticed the airspeed indicator was pegged at the limit of 1,100 kilometers per hour: about 680 miles per hour. As his nose pitched sharply down, Mutke could no longer control the 262, later recalling to American aviation historian Walter J. Boyne: "I moved the stick wildly around the cockpit. For a brief moment, the airplane responded to controls again momentarily, then went back out of control. The plane still did not respond to pressure on the stick so I changed the incidence of the tailplane. The speed dropped, the aircraft stopped shaking, and I regained control."

By manually altering the angle of the horizontal tail, which could be done from the cockpit, Mutke disturbed the airflow over his stabilizers and slowed from supersonic to subsonic speed. When this occurred, the shock wave generated by supersonic air moved forward, allowing the nose to lift again and control to be restored. Mutke was able to throttle back, which flamed out his engines, then slow to a controllable 300 miles per hour. Managing to land, he and the ground crew discovered popped rivets and warped wings.

Limited to 0.86 Mach by technical order, it was physically *possible* for the Me 262 to exceed the speed of sound in any sort of dive. At full throttle in the thin air above 30,000 feet and diving, this becomes more likely if there is no structural failure. The damage to Mutke's jet indicates that any sort of prolonged flight under those conditions would have been disastrous. A computer-modeling simulation conducted at the Technische Universität München in 1999 suggested it could be done—but a simulation is not the real thing.

Did it happen?

Ken Chilstrom, one of the original USAF test pilots who flew a captured 262, doesn't think so. "The engines could have done it," he stated in a 2017 interview for this book, "but structurally I don't think the plane would have held together."

"This could have just been the jet I flew though," he added. "German quality was very problematic at that phase of the war, and each jet was put together a bit differently."

No one knows, and to date nothing has been found in Messerschmitt's surviving wind tunnel data to indicate it had ever been tried. Yet given the combat accidents suffered by Allied fighter pilots who lost control during high-speed dives, transonic speeds had clearly been reached. And if straight-winged, piston-engined aircraft could cross into that region, then a swept-winged jet certainly could so.

How then did mankind, who had only truly flown forty-two years earlier, progress to the point where flight past the speed of sound was possible? For even if it did not occur on that March day in 1945, it was certainly possible, and extreme velocity was the way of the future. This was a hard lesson learned on both sides as jet-powered Me 262's operated far beyond the capabilities of Allied piston-engined fighters and defensive guns from bomber formations. Fortunately, there were too few of them to make a significant impact, but this same German technology, inherited by the Allied victors, would live on to shape global geopolitics during the late 1940s and early 50s.

With the atomic bombing of Hiroshima and Nagasaki at the end of World War II, the nuclear age had begun. War planners now faced a radically new calculus: the ability to get a bomb over target faster than one's enemy trumped "conventional" tactics. Meanwhile, a new conflict, the Cold War, was emerging. Tensions between the United States and the Soviet Union were already high as the fragile, wartime alliance frayed to a point where, in 1946, Joseph Stalin openly declared the coexistence of capitalism and socialism to be impossible. Justifiably wary of challenging the American military, Soviet engineers and scientists raced to close the technical gap. Jet aircraft, specifically long-range strategic bombers capable of delivering nuclear weapons, were initially viewed as the best option for power projection. This, Moscow believed, would permit the intimidation possible for multi-hemispheric, Communist expansion and, if the situation deteriorated, would provide a Soviet capability to wage open war with the United States.

. Equally cognizant of the military and ideological threat posed by the USSR, American leaders were determined to maintain their technological edge and therefore blunt Communist ambitions. If atomic or thermonuclear warheads could be delivered onto an enemy's soil, as had been done against the Japanese, then who would risk armed confrontation with the United States?

Even as they returned home to try to enjoy a world without war, a small group of incredibly daring American fighter pilots were poised to shake the foundations of man's achievements to their core; to kick open the door of modern air combat tactics and change the world as it currently existed. Having survived combat and vanquished one set of threats these men knew other evils would rapidly emerge to fill the vacuum, and so they did. Ideological and nuclear threats capable of a mass destruction unthinkable a decade earlier, and for a few short years the safety of the world was dependent on an extremely narrow margin defined and redefined by military technology; an anthropogenic Pandora's Box that could either be contained by the West, or opened by the East through the possession of a single key: speed.

In the years immediately following World War II speed was the yardstick by which aircraft, the "long rifle" of the modern age, were measured. Before the development of intercontinental or submarine-launched ballistic missiles, jet aircraft capable of high subsonic flight were essential to the ambitions of both superpowers, and to the making of war or the preservation of peace. The ability to fly farther, longer, and, above all, faster, could either balance or tip the uneasy strategic equilibrium of the postwar era. Manned supersonic flight was the prize as it offered an area of exclusion—a profound combat advantage—for those who possessed it against those who did not. This quest began with experimental aircraft and fighters, for a supersonic interceptor could destroy a bomber well before it could drop its lethal, thermonuclear payload. Defensive weapons, coupled with radar guidance, had not yet been developed to counter such speed, so it was believed that such combat operations could be conducted with relative impunity, and thereby ensure victory or maintain the peace.

Men like Ken Chilstrom, George Welch, Chalmers Goodlin, and Chuck Yeager had already risked everything for their country during the Second World War; their own lives had been disrupted and forever altered, yet they did not hesitate to roll the dice again in the name of duty, danger, and honor. These were hardened men born in the turmoil following the Great War, raised to maturity during the chaos of the 1930s, and honed sharp by combat. They understood, if others did not, that peace not backed by strength was an empty hope. These men were part of an America that accepted, albeit reluctantly, a mantle of global leadership that could not be discarded if the world were to remain safe for their children's children. Just as it never occurred to them to dodge their wartime obligations, they could not, and would not, shrink from a new challenge. Only pilots such as this could chase the demon beyond whatever barrier existed, and finally take mankind faster than the speed of sound.

Part One

—————

ORIGINS

".. . and if you gaze into the abyss,
the abyss gazes also into you."

FRIEDRICH NIETZSCHE

One

Flying Monks to Mud Ducks

ong before it was conjured by men in aircraft, the demon's presence was known. Whips, and their corresponding sonic booms, had been in use at least since the Egyptian Middle Kingdom, some 4,000 years past. China, the Indian Maurya Empire, and ancient Rome all used whips in some form or another and heard, although without comprehension, the distinctive crack of the sound barrier being broken at every stroke. Closer to our own time, fifteenth- and sixteenth-century physicists were aware that the speed of sound had a limit and were determined to define it.*

These were individuals like Sir Isaac Newton, who, in his seminal *Principia Mathematica* of 1687, calculated this limit at 979 feet per second but failed to account for the influence

---

* Those interested in the scientific underpinnings of mechanical flight should review the appendix, "Aerodynamics 101," on page 275.

of heat. This error, incidentally, was corrected the following century by Pierre-Simon Laplace. Another Frenchman, Marin Mersenne, calculated the illusive number at 1,380 Parisian feet per second while Robert Boyle, of Boyle's law fame, arrived at 1,125 Parisian feet per second.* William Derham, a clergyman in Essex, England, came the closest in 1709 at 1,072 Parisian feet per second. Derham had friends fire shotguns from known locations and he painstakingly measured how long it took for sound to travel to his position. Early flintlock firearms, at least as far back as the fourteenth century, were also capable of producing supersonic projectiles.†

The idea of an aircraft flying faster than sound was considered a very real possibility within two decades of Wilbur Wright's 1904 Kitty Hawk flight. United States Navy Lieutenant Commander Albert Cushing Read, first to fly from America's East Coast to Europe in 1919, declared that he could see the day when "it will soon be possible to drive an airplane around the world at a height of 60,000 feet and 1,000 miles per hour." Though belittled at the time, Read and others *knew* it could happen once technology caught up with vision. This is often surprising to those of us raised with aircraft and air travel as we consider this capability a modern invention; and it is, from a practical standpoint, yet there have been significant, albeit oft-forgotten, aviation milestones stretching back for over 1,000 years.

Man had likely been fascinated by flight from the beginning. One can imagine a heavily muscled, low-browed *Homo erec-*

* A Parisian foot is 1.0657 times the standard U.S. measurement.

† 1,426 fps for an American .69-caliber flintlock with a 120-grain load utilizing a 412-grain ball.

tus peering uncomprehendingly at a swooping bird and perhaps dimly wondering why he could not do the same. Jumping from cliffs, later leaping from towers or attaching oneself to a kite—all these efforts to fly, and doubtless many more, had been attempted over the centuries. Though usually resulting in a painful death or lifelong injury, there were some successes nevertheless.

Though the Chinese had been flying kites since the tenth century BCE, the first recorded human flight is generally credited to Abbas ibn Firnas, a Berber polymath living in Moorish Cordoba. After studying kites, and all the known previous attempts at gliding, Ibn Firnas constructed a light, wooden wing, very much like a modern hang glider, from silk and feathers. In 875 CE, at the age of seventy, he jumped from Jabal al-Arus (the Bride's Hill) outside Cordoba and, by some accounts, glided for ten minutes over the Guadalquivir River valley. It was a successful flight, though Ibn Firnas had neglected to consider the problem of landing and he was badly injured.

In the early eleventh century Eilmer of Malmesbury, a Benedictine monk from southwest England, built a wing from cloth and wood then leaped from the West Tower of his abbey. Gliding downhill against the wind for at least fifteen seconds, he traveled over 200 yards. Unfortunately, as with Ibn Firnas, the monk had not considered the finer points of flying, in this case, control. His apparatus had no tail and was out of balance, so when the wind changed, Eilmer came crashing down, breaking both legs.* Other men followed. All were enamored of

---

* A stained-glass window in the vestry of Malmesbury Abbey shows Eilmer and his "glider."

flight, were courageous, and had no real idea of what they were truly attempting. Broken legs and necks were common. By the end of the seventeenth century the Italian physiologist and biomechanical pioneer Giovanni Alfonso Borelli definitively concluded in his *de Motu Animalium* that humans lacked the musculature to sustain flight by flapping wings.

On a moonlit summer night in 1793 a man named Diego Marín Aguilera jumped from Coruña del Conde castle in northern Spain. Soaring at least 1,000 feet across the Arandilla River, his wooden machine very likely caught an updraft from the valley floor and the stress fractured a metal joint. Crashing near Heras, he narrowly escaped being burned as a heretic by the town's inhabitants, who believed a flying human was an affront to God.

These men, if the accounts are accurate, had at least progressed from near suicide to basic flight. To be sure, gliding can be considered a form of flight, as at its best a glider can have a sustained time aloft and be *controlled* by a pilot. Elementary as this sounds, it took centuries of trial, error, and death to progress past this basic point. True flight, as we shall see, must have a power source able to propel an aircraft through the air with sufficient velocity to produce lift. The bird or bat has its wings, but man must have something else.

Yet the scientific achievements of very early aviation pioneers are often underestimated or overlooked entirely, and this is quite unfair since we acknowledge debts owed to visionaries from other fields. Antonie van Leeuwenhoek revealed the interior structure of cells in the seventeenth century while Louis Pasteur's contributions to bacteriology and Gregor Mendel's to genetics both followed during the next century. Tycho Brahe discovered a supernova in 1572, while Galileo found craters on

our moon and identified the Milky Way galaxy. Danish astronomer Ole Rømer correctly measured the speed of light by 1675,
and the Royal Society published Ben Franklin's *Experiments
and Observations on Electricity* in 1751.

Engineering ceilings of all kinds were shattered during the
Industrial Revolution, so it should be no surprise that aerodynamics advanced as well. If Boston could implement the first
municipal electric fire alarm in 1852, or Thomas Edison could
erect the first dedicated research and development laboratory
at Menlo Park in 1876, then Francis Wenham's wind tunnel
should be no less revered. Hydroelectric power plants, commercial electrification, and even experimentations with millimeter
wave communications were all conducted in the late nineteenth
century, therefore Horatio Phillips's cambered airfoil, or Félix
du Temple's first sustained flight by a true heavier-than-air machine in 1874 ought to be as well known—yet they are not.

Perhaps one reason lies with the mystique surrounding
flight. Unlike steam power or electricity, flying was not an ac-
tivity that benefited the masses until well into the twentieth
century, so it largely remained the province of the scientific
community or the independently wealthy. Then, less than two
weeks after the 1773 Boston Tea Party, a man was born in Yorkshire, Great Britain, who would arguably usher in the modern
age of aviation. George Cayley, a self-educated baronet and a
man of indefatigable imagination, designed caterpillar tractors
and artificial limbs before studying avian physiology to aid his
understanding of his true passion: flight.

In 1799 Cayley was the first aerodynamicist to break the process of flight apart into the distinct components of lift, weight,
thrust, and drag. He insisted that thrust, or some manner of
propulsion, was an independent factor that must be practically

solved for man to truly fly. Cayley also correctly envisioned the modern structure of an aircraft with a fuselage, forward wings, and a cruciform tail surface. By 1804 he had constructed a flyable model glider, and five years later his three-part essay "On Aerial Navigation" was published in *Nicholson's Journal of Natural Philosophy, Chemistry and the Arts*.

Sir George understood about centers of gravity, and that it was the pressure differential acting on an airfoil that generated lift. Above all, Cayley realized that, unlike a bird, a man must generate lift through a separate form of propulsion. Steam would not suffice; the engine was inefficient and entirely too heavy. Internal combustion, he felt, was the only realistic solution and Cayley spent a good deal of his life theorizing about just such an engine.

He flew models and developed full-scale gliders, including one flown by a ten-year-old boy in 1849. By 1853 he had constructed a craft that could remain airborne, with his unenthusiastic coachman as pilot, for about 500 yards. Upon landing, the unhappy servant told Sir George that "I was hired to drive, not to fly," and he promptly gave notice.

However, it was Cayley's methodical evaluation of his concepts that opened the doors to purposeful, *systematic* testing. With a whirling-arm device used to design windmill blades, he added a paper airfoil and adapted the contraption for surprisingly accurate studies of lift. Dr. John Anderson, the preeminent aerodynamicist of our time, writes that Cayley's measurements were "accurate to within 10 percent based on modern aerodynamic calculations." Toward the end of his long and interesting life, Cayley summarized his work by formulating the essence of all modern aircraft within the simple, but as yet undefined principles of lift, propulsion, and control.

With the doorway to flying now framed, the subsequent century of flight research and development was largely a stair-step progression of ideas, techniques, and revelations. There was some cooperation, much jealousy, and often open disdain among the competing worlds of academia, theoretical engineering, and those physically attempting to fly. Yet without an efficient means of propulsion, much of this initial progress necessarily centered on gliders.

With theory and practice warring with each other, the focus shifted as various problems were addressed and eventually solved. By the late eighteenth century, lift was well understood so the emphasis moved to creating thrust and mastering control. If you recall, Cayley pointedly separated propulsion from lift and this was a crucial point. Flying requires thrust, whether it is self-generated like a bird, bat, and insect, or via some type of artificial propulsion such as an engine. If you are not flying under power, then you are not flying; you are gliding or, even worse, you are floating.

Early experiments in flying sought to emulate birds, which was reasonable enough as they were the most obvious examples of successful flight, yet it was impractical. Birds are able to fly due to a combination of evolutionary advantages, such as honeycombed bones that yield a very strong, yet extremely light frame. This frame is covered with keratin feathers that are molded, or preened, into highly efficient airfoils capable of producing lift. But a bird, like a man, still needs to generate thrust in order to produce lift. In the bird's case, this is possible due to a high metabolism that enables its muscles to work more than twice as fast as other mammals. This permits flapping that generates enough thrust to get air moving over the wings, which in turn produces lift.

Once it was understood that man could not replicate these natural advantages, then artificial methods of generating thrust were explored, and the quest for powered flight moved forward. The results speak for themselves with the nineteenth century witnessing the advance of theoretical aerodynamics into workable flying machines. Some of these, like William Henson and John Stringfellow's aerial steam carriage (also called the Aeriel), were wildly impractical; how could a 30-horsepower engine propel a machine weighing well over a ton? Indeed, its 150-foot wingspan and gigantic 4,500-square-foot wing area exceeded that of a modern Airbus 320.*

It never flew, of course, but was nonetheless influential by inspiring others through its form and potential. Dr. Anderson says of Henson's monstrosity, "Here is an excellent example of the still technically undeveloped state of the art of airplane design in the first part of the 19th century." Yet Henson's machine also seemed to graphically illustrate the rather profound differences between aerodynamic theorists, academicians, and the designers of aircraft. One problem was to separate flight from propulsion—and no one had a really clear idea how either worked.

Clément Ader, on the other hand, actually *did* get a machine airborne under its own power: a 20-horsepower steam engine. A French electrical engineer, Ader specifically looked to nature for inspiration and by 1890 completed a machine he named the Éole. On October 9, the bat-winged contraption staggered into the air near Armainvilliers and managed to remain aloft for 165 feet. Though this event was a startling aviation first, a

---

* The A320 wingspan is 117.4 feet with a wing area of 1317.5 square feet.

manned craft flying under its own power, it still did not qualify as a "flight" since Ader had no way to control, or physically "fly," the aircraft.

Neither did Hiram Maxim. Arrogant and vain, Maxim was unquestionably brilliant, and behind his unpleasant façade lay a first-class brain coupled to a fertile imagination. A self-educated inventor, he patented the original machine gun in 1883 and incorporated the Maxim Gun Company the following year. After emigrating from America to the United Kingdom, his wealth permitted the freedom to pursue other interests, including aviation. When asked if he could build a flying machine, Sir Hiram replied, "the domestic goose is able to fly and why should man not be able to do as well as the goose."

Methodical and precise, he was the first aviation pioneer to derive specific wind tunnel data toward a specific design. Like Ader, Maxim's immediate goal was to get a manned aircraft aloft under its own power, so he leased Baldwyns Park outside London, and built a hangar to accommodate his project. The result was a *four-ton* flying machine powered by a 362-horsepower steam engine that would propel the craft down 500 yards of railway track. Maxim mounted extra raised wheels on his apparatus that would catch a wooden safety rail running parallel with the track. This, he reasoned, would keep the machine from getting more than a few feet above the ground and prevent crashes.

On July 31, 1894, he did just that.

Under full power the three-man crew reached 42 miles per hour, and the giant seventeen-foot, ten-inch propeller kept the craft airborne (at two feet) for over 300 yards. Yet for all his considerable talents Maxim, like many others, could not conceive of the aircraft in more than diversionary terms, an engineering challenge. "But I do not think," Maxim once stated, "the flying machine will ever be used for ordinary traffic and for what may be called 'popular' purposes. People who write about the conditions under which the business and pleasure of the world will be carried on in another hundred years generally make flying machines take the place of railways and steamers, but that such will ever be the case I very much doubt."

But since Maxim and Ader succeeded in getting into the air under their own power, why did they not get credit for the first flight? Obviously a few basics were understood, at least as far as building an airfoil that produced sufficient lift to overcome weight and get airborne. Maxim's machine was powerful enough to generate a very respectable unit of horsepower for each twenty-two pounds of weight, and Ader's subsequent designs were quite similar.

In the end, this comes down to how true flight is defined. Whereas getting airborne under power is quite different from gliding, so too is piloting your aircraft as you choose once aloft. When inventors, engineers, and others expanded on Cayley's separation of lift, thrust, and drag, a final component was eventually realized: control. Ader and Maxim did produce thrust, which in turn generated lift, so in this respect they were definitely a bridge between the world of gliding and that of true flight. To truly fly, one must have control of the aircraft. To "feel" the plane and adapt to the continuously changing circumstances around it.

In other words, to be a pilot.

By the close of the nineteenth century those most successful in aircraft development were harnessing the theoretical aspects of the new science with the ability to conduct the experiments themselves, to fly their own machine. From this point of view

Otto Lilienthal was arguably the first test pilot in the modern sense of the title. A mechanical engineer by training, he believed that each component of flight—lift, propulsion, and control—had to be fully understood and the issues with each solved to arrive at a comprehensive solution.

Using practical, engineering-based processes, Lilienthal was specifically concerned with the variations in air pressure on a wing resulting from changes in the angle of attack. He systematically measured this, and other hypotheses, during some 2,000 flights in sixteen types of gliders near his home in Steglitz, or his testing area over the Rhinow Mountains. He even constructed a small hill in Lichterfelde near Berlin so he could always launch himself into the wind. A monument was constructed on the site of Lilienthal's research shed in 1932 and it is there still, a delightfully Germanic Stonehenge surrounding a stone globe that overlooks a rectangular pond.

Perhaps Lilienthal's greatest contribution was the formulation of aerodynamic coefficients that permitted the use of dimensionless quantities to characterize forces acting on an airfoil. This greatly simplified lift and drag calculations and permitted progression into modern aerodynamic design. Like Horatio Phillips, Lilienthal arrived at the conclusion that cambered airfoils were a necessity for an effective wing. Interestingly, this was done independently, so Lilienthal was unaware of any competing work until he filed a patent application in 1889 and discovered it had already been granted to the Englishman. That same year he also published *Birdflight as the Basis of Aviation*, a compendium of verified aerodynamic data that included results from his own experiments and the seminal work on flight.

Yet despite his visionary efforts, Otto Lilienthal suffered

from the rather serious delusion that the ideal solution for powered flight would be an ornithopter; that is, a machine that flies by flapping mechanical, rather than static wings, which generate lift while being propelled through the air. Unlike Ibn Firnas and Eilmer the monk, Lilienthal was aware that a man could not produce sufficient muscular force to sustain flight by flapping since our bodies are too heavy relative to the muscular force produced, and we have the wrong type of muscles. He actually constructed a one-cylinder engine to flap his glider's wings and commenced testing in Berlin during the spring of 1894. Having absolutely no success with this, he returned to gliders with hopes of producing them commercially for sport.

In common with his predecessors, Lilienthal had looked to birds for answers and this partially explains his ornithopter fixation. In any event, as his gliders had no control surfaces he, like the birds, relied on shifting his own weight to maintain altitude and direction. On a sunny Sunday afternoon in August 1896 he caught an updraft and the glider stalled, sending Lilienthal into a fifty-foot fall that broke his back. He died the following day, a stark reminder that with no control, lift is a force that can kill.

In concert with Lilienthal, Octave Chanute believed in stable aircraft and devoted his considerable expertise in improving structural designs. A native-born Frenchman who became a U.S. citizen at age twenty-two, Chanute gained early fame as an engineer and urban planner. Designing both the Kansas City and Chicago stockyards, he was also the chief engineer for the Chicago & Alton Railroad. On July 3, 1869, the thirty-seven-year-old Chanute's Hannibal Bridge opened in Kansas City, a tribute to his structural engineering skill and adaptability, two qualities that would propel him to the forefront of aviation.*

Always attracted by a challenge, in his midforties Chanute set out to overcome the technical difficulties plaguing aircraft enthusiasts, partnering with Augustus Moore Herring. The pair eventually constructed a lightweight biplane with extremely strong, straight wings. He ingeniously adapted the Pratt bridge truss design, which utilized a combination of vertical and diagonal members and evenly spread the aerodynamic load. This was a deliberate and highly significant departure from previous wings patterned after birds or bats. Chanute was aware that for man to fly he needed an engine for propulsion, and existing aircraft frames were either overengineered, like Maxim's monstrosity, or, as with Ader's Éole, too frail to support heavier equipment.

On May 9, 1896, a man named Samuel Pierpont Langley proved that powered flight *was* possible with his Langley Aerodrome Number 5. Catapulted from atop a houseboat on the Potomac River, it managed to "fly" about thirty-five feet under its own power. The cambered, tandem wings spanned a bit over thirteen feet but had an unfortunate tendency to flex once launched, which, of course, altered the craft's aerodynamic properties. Like Lilienthal, Langley was fixated on the physical aspects of getting a craft airborne so, as with his predecessors, he was uninterested in controlling a machine—he just wanted to get it airborne.

A physicist and astronomer by education and training,

---

* Illinois's Chanute Field, later Chanute Air Force Base, was named in his honor.

Langley was quite capable of complex calculations and he applied this knowledge to his newfound aerodynamic interests. His Power law, which essentially stated that a faster aircraft required less power to sustain speed than one flying slower, was immediately controversial and rejected by such luminaries as the Wright brothers and Otto Lilienthal. In fact, Langley was halfway correct. What is true, and he was decades ahead of his time in seeing this, was that a "fast" wing has a lower angle of attack and therefore drag is considerably less. Less drag means less power is required just as greater drag caused by a "slow" wing with a higher angle of attack requires more power to push it through the air. This is the "back side" of the power curve, sort of an aerodynamic point of no return. What he got wrong, because his apparatus was incapable of producing it, was that at velocities exceeding 72 feet per second this reverses.

His acquaintance with Assistant Secretary of the Navy Theodore Roosevelt and the onset of the Spanish-American War in 1898 provided Langley with a princely $50,000 grant from the U.S. Army Board of Ordnance and Fortification. He was to design, construct, and produce a full-sized aircraft capable of flight with a pilot aboard so, with the stroke of a pen, Samuel Langley became the first aviation defense industry contractor. Skeptics abounded, but so did Langley's optimism and five years later, on October 7, 1903, his "Great Aerodrome" was ready to fly.* Charles Manly, Langley's assistant and pilot, started up the 52.4-horsepower internal combustion Balzer-

---

* The name was a mistranslation of ancient Greek. *Aerodrome* means a place from which aircraft are flown, not an aircraft itself, as Langley believed.

Manly engine, smiled, and waited atop the houseboat for the signal. The nearby tugs gave a few horn blasts, and his mechanic cut a cable that launched the aircraft. A watching reporter from the ·Washington Post wrote:

> There was a roaring, grinding nose—and the Langley airship tumbled over the edge of the houseboat and disappeared in the river, sixteen feet below. It simply slid into the water like a handful of mortar.

Langley tried again two months later. With ice on the Potomac and a cold wind blowing, they launched at 4:45 on a cold, windy afternoon. This time the Aerodrome's wings snapped and Manley once again ended up in the river. A congressman's sarcastic comment, oft quoted by the press, named the Aerodrome a "mud duck which will not fly fifty feet." Ridiculed and shamed, Langley quit and died, discouraged and brokenhearted, in 1906.

Still, he had accomplished what he had intended: a successful powered flight by a heavier-than-air machine. True, it was of short duration, unmanned, and uncontrolled; but that was coming in December 1903, with two obscure men from Ohio who captured immortality on a bleak, cold North Carolina beach.

They were inseparable brothers and lifelong bachelors with rudimentary high school educations. They certainly lacked Samuel Langley's scientific training, Lilienthal's and Ader's engineering background, and Cayley's imagination; but Orville and Wilbur Wright grasped the essential and previously minimized aspect of *control*. It was, the brothers recognized, the final basic problem to be solved. They knew that without a pilot's control of his powered, heavier-than-air craft there was no true human flight.

Born into the sturdy, respectable middle-class family of Bishop Milton Wright, Wilbur and Orville seemed destined to follow their father into the church and business, respectively. They founded several newspapers, the *West Side News* and the *Dayton Tattler*, followed by the famous Wright Cycle Exchange on Dayton's West Third Street in 1893. Lilienthal's death in 1896 was largely responsible for attracting the Wrights to aviation in that it presented formidable challenges in several areas and was an endeavor as yet unconquered.

By this time, much was known and understood about basic aerodynamics. Cambered wings, lift and drag, wind tunnels, and, through Maxim and Langley, proof that an aircraft could physically get off the ground under its own power. Yet the deaths of Percy Pilcher, Lilienthal, and others convinced the brothers that flying would never be safe, and therefore never accepted, until it could be satisfactorily controlled. With this in mind they set themselves to the task of defeating this final, elusive obstacle to manned flight. Besides a natural aptitude for science and practical engineering, the Wrights had the tremendous advantage of decades of research and experimentation to draw upon, which they did quite analytically and methodically.

As with their predecessors, the brothers began by studying birds and noticed that directional control came from a twisting of their wingtips. This altered lift over each wing and produced a rolling motion to "bank," or turn, the animal at will. The discovery was crucial and with their experience in cycling seemed perfectly logical. James Howard Means, editor of the influential *Aeronautical Annual*, would opine in 1896 that:

The slow development of the flying machine in its early stages finds its analogy in that of the bicycle. The machine has been improved very gradually; most of the modifications have been slight; yet some of the stages have been marked with great distinction.

A workable method of control was absolutely one of these stages, and the Wrights' solution was termed "wing warping." The story is told that Wilbur, while twisting an empty cardboard inner tube box one day at the bike shop, noted that when one edge went down the other came up. If, he thought, this could be replicated mechanically on his aircraft wing, then lateral control could be achieved—just like a bird. The brothers found by removing the diagonal fore-aft bracing wires at each end there was enough flexibility in the wingtips to twist, or warp, them at will. Running the span-wise wires through a hip cradle enabled a prone pilot, by shifting his weight, to laterally control the craft. Wing-warping tests performed with their 1900 glider were entirely satisfactory.*

They would spend the next two years traveling back and forth between Ohio and North Carolina to test and validate their innovations. Often discouraged, they stubbornly persevered and incorporated each improvement into their glider designs. The Wrights discovered that although there were prodigious amounts of previous work to consult, a lot of it was incorrect or, in the case of Lilienthal's lift table, they were

---

* In fact, the wing-warping mechanism was included in the Wright Flying Machine patent filed on March 23, 1903. No. 821,393 would be granted on May 22, 1906.

applying it incorrectly.* This would gradually lead to the revelation that while there were absolutes in aerodynamics, each
aircraft design would dictate how those absolutes were to be
applied. For Wilbur and Orville this meant discarding much
preceding technical work, as Lilienthal had done, and con-
structing their own wind tunnel *and* custom instruments.
The wind tunnel was 6 feet long with a 16-inch cross section
and the fan, rotated by a 1-horsepower gasoline engine, could
generate a 30 mph wind stream. They added a glass observation window to observe, in real time, the efficiency of their
tests.

This was a logical step for them to take, yet decidedly marked
the entry of aviation science into the modern age where all the
data, theories, and ground experimentation used to build the
aircraft are then validated by that aircraft, and its pilot. In other
words, test flying. Hiram Maxim had been the first to do this,
to a degree, but his aims were limited to physically getting a
powered craft off the ground.

The Wright brothers intended *to fly.*

And so they did.

By December 1903, just after Samuel Langley had given up
on his Great Aerodrome, Orville and Wilbur solved their longitudinal and lateral control issues and were ready to take their
newly christened Wright Flyer into the air. On December 14,
the brothers flipped a coin and Wilbur won the toss. Perched
atop the dunes at Kill Devil Hills, the aircraft was fixed to a rail
and angled slightly downward. Starting the engine, a 12-horse-

---

* See John D. Anderson's superb analysis of this in *The Airplane: A History
of Its Technology*, pp. 102–104.

power, gas-powered, four-cylinder design of their own, Wilbur raced down the incline and into the air.

Overpulling, he got the nose too high, stalled, and subsequently crashed, causing enough damage for three days of repairs, but with no injury to himself. December 17 dawned with a cold, gusty wind blowing over the sand. At 10:30 A.M., Orville Wright shook hands with his brother, started the engine, and stared at the dunes toward either death or immortality. Releasing the restraining line at 10:30, the aircraft puttered down the rail into a 27 mph headwind with Wilbur running alongside holding one wing for balance. Suddenly, after a short forty feet, the Flyer wobbled into the air and the volunteers gathered along the beach began cheering.

Twelve seconds and 120 feet later Orville touched down after completing the first manned, *controlled* flight of a heavier-than-air craft under its own power. Ecstatic, the brothers swapped places for two more flights and at noon, with Wilbur at the controls, the Flyer remained airborne for 59 seconds and covered an astonishing 852 feet. One of the volunteers summed up the event, and man's true entrance into aviation, by shouting, "They did it! They did it! Damned if they didn't fly!"

On that Thursday morning at Kitty Hawk, Orville and Wilbur Wright conquered the air with little comprehension of how far, how high, and how fast their accomplishment would take mankind. It opened the door to a new world that has proven time and again that there is, and very likely always will be, another challenge waiting in the thin air beyond the clouds.

Two

======

The Cauldron

To a large extent we are kites in the wind with regard to fate. Governments rise and fall, fads come and go, technology soars, trends wax and wane, and most of it seems beyond our control. But is this really true? Do we make our times or do our times make us? Surely, this is an enormously complex question, yet the quest to conquer the speed demon was accelerated, if not created outright, due to the pivotal, cataclysmic upheavals that exploded on the world during the first half of the twentieth century.

Humans rarely change, and when they do, it is not a rapid transformation. To a large degree then, it is the times and their events that create the people needed to face the unique situations of each era. This means, given the necessity, we would rise up and meet challenges today just as our ancestors did before us. True, other empires had risen and fallen; evil had battled good equally unambiguously, and technological advancements had spiked before, yet the 1940s were different. Man had created weapons that could obliterate entire cities, he could freely move beneath the oceans, and unquestionably man now ruled the skies. Ken Chilstrom, George Welch, Chuck Yeager, Bob Hoover, and Chalmers Goodlin were all part of this; they were born following one great disaster, grew up in another, and came of age during the most horrific war in history. These men were all combat fighter pilots—Welch and Yeager would become aces—and all would enter the rarified world of test flying following World War II. Though they had much in common, they faced the demons of life, war, and flying in very different ways. So what factors and influences in particular molded them? How did they become who they were, and what made it possible for them to chase the demon past the speed of sound, pulling mankind into the supersonic, nuclear age?

George Welch was ten days old on May 28, 1918, when the American Expeditionary Force launched its initial offensive action and America's first victory in the Great War.* The consequences of that battle and that war changed his life, and our lives as well. On November 11, 1918, the armistice was signed and the several million U.S. soldiers in France began shipping home to their families, their former jobs, and the lives they left behind. The government, with no forethought whatsoever, abruptly canceled most of the war contracts that had produced America's booming economy. Jobs vanished overnight, and

---

* The 1st Battalion, 26th Infantry, of the U.S. 1st Division was led by Major Theodore Roosevelt Jr.; he was so concerned for the welfare of his soldiers that, at his own expense, Roosevelt purchased new boots for every man. Returning to military service for World War II, Roosevelt was the oldest man at fifty-six (and the only general officer) to come ashore by sea with the first wave of the invasion.

returning veterans wanted those that remained. A recession ensued and, exacerbated by race riots and fears of immigrants and anarchists, the nation plunged into a decade of profound uncertainty and social changes.

Ken Chilstrom was born during all this on April 20, 1921, in a tiny town called Zumbrota, on the north fork of the Zumbro River in southeast Minnesota. It was farming country, predominantly Lutheran, heavily conservative, and like most childhood experiences it left a permanent mark. "What I learned about farmers and the land taught me discipline and responsibility," he recalls. Discipline and responsibility. Two words that would define the man for all of his long, exciting life. Born to second-generation Swedish immigrants, Ken took after his mother, Emma, a schoolteacher, but he greatly admired his father, John, who ran a general store. "My father was such a good man in so many ways. I never heard him swear or use bad language."

By the time Ken's father moved the family to Hartford, Wisconsin, Warren G. Harding had become president, Edgar Rice Burroughs released *Tarzan the Terrible*, and jazz appeared in New Orleans. Both the Eighteenth Amendment and the Volstead Act, otherwise known as the National Prohibition Act, had gone into effect so America was legally dry. The Nineteenth Amendment, granting female suffrage, had also passed and proclaimed "the right of citizens of the United States to vote shall not be denied or abridged by the United States or by any State on account of sex." Benton MacKaye would propose the Appalachian Trail, and the first Miss America Pageant was held during September in Atlantic City, New Jersey.

Three months after Bob Hoover's January 1922 birth in Nashville, the lid blew off the Teapot Dome scandal. America was alternately shocked and fascinated as the government's corruption and incompetence was exposed and President Harding publicly humiliated. Secretary of the Interior Albert Bacon Fall used his position to secretly sell a lease to the Mammoth Oil Company for Wyoming's Teapot Dome, officially known as U.S. Naval Oil Reserve Number Three. In return, Fall received $260,000 in Liberty bonds and at least $100,000 in cash.

But the news wasn't all glum.

In May 1922 construction began on Yankee Stadium and Washington, D.C., witnessed the dedication of the Lincoln Memorial. By 1923 the country's fortunes were changing for the better. Harding died in office and was succeeded by his dour vice president, Calvin Coolidge of Vermont. The recession had faded, and though the next seven years would test America's respect for government, politics, and religion, there was room for optimism. Refrigerated shipping made it possible to obtain a wide variety of fresh food year-round, and the virtues of vitamins had been discovered. *TIME* magazine hit the streets, and the first Winter Olympic Games were held in Chamonix, France, with the United States picking up four medals, including the gold for the Men's 500 Meter Speed Skating.[*] Nineteen twenty-three also saw the births of Chalmers Hubert Goodlin, later known as "Slick," and Charles Elwood "Chuck" Yeager.

Amid the ongoing strife of the '20s, especially the "red scare" of Russian Bolshevism and ongoing conflicts between faith and science, flying was a positive, exciting influence. There were others, to be sure, and much of the decade was certainly not grim. Fashion had changed drastically, with women show-

---

[*] Additionally two silvers were won for the Ladies Single Figure Skating and Men's Ice Hockey, respectively, and a Bronze for Ski Jumping.

ing more skin than ever before and there was the excitement, at least within larger cities, from speakeasies, illicit drinking, and new dances that encouraged close contact in dark, smoky places. Over 800 movies were made each year, and folks routinely saw "pictures" several times each week in stupendous new theaters like San Diego's Balboa, the Saenger in New Orleans, or the opulent 3,353-seat Kodak Hall in Rochester, New York.

Then there was aviation.

The world of flight was a source of pride, inspiration, wonder, and, as it still remains today for many, a bit of a mystery. The decade got off to a tremendous start with newsmaking, eye-popping events during the summer of 1919. A trio of U.S. Navy Curtiss flying boats ponderously lifted off from Naval Air Station (NAS) Rockaway on Long Island during the morning of May 8, 1919. They turned northeast and headed up the North American coast for Trepassey, Newfoundland. All three later departed Newfoundland for Horta, in the Azores, assisted by Navy warships stationed at fifty-mile intervals with illuminated spotlights and flares to show the way. Eventually one of the planes, an NC-4 flown by Albert Cushing Read, landed at Plymouth, England, on the last day of May, via Portugal and Spain. The public was enthralled; the Atlantic Ocean had been crossed from continent to continent.

In June, Captain John Alcock and Lieutenant Arthur Whitten-Brown of the Royal Air Force took off from Lester's Field outside St. John's, Newfoundland, heading for the United Kingdom, some 1,900 miles to the east. In an open cockpit Vimy bomber, they flew all night through snow, fog, and ice where finally, sighting the Irish coast fifteen hours later, they landed by mistake on Galway's Derrygimla Moor. This was the first nonstop, heavier-than-air flight from North America to Europe, and it captivated the world as did the transatlantic crossing by a British airship in July. Following a 108-hour, 12-minute passage, the U.S. Naval observer aboard, Lieutenant Commander Zach Lansdowne, parachuted onto American soil then personally moored R-34 at Roosevelt Field, on Long Island's Hempstead Plains.* The mystique of aviation had indeed captured the world's imagination.

This fascination for all things flying was greatly magnified by the romantic, but somewhat misplaced, notions surrounding combat aviation during the Great War. American boys like Charles Lindbergh and Jimmy Stewart thrilled to stories, real or exaggerated, about Mick Mannock, the Red Baron, and Eddie Rickenbacker.† In four years the war had transformed aviation from a fad, a sporting curiosity, to a serious, tactical weapon. This led to more powerful engines and better designs, and prolific innovations in all other aspects of aerial warfare raced forward as both sides continuously designed their way out of combat shortcomings.

Tough and accurate machine guns such as the Spandau and Vickers were manufactured; synchronization gear was perfected that permitted continuous machine gun firing through a propeller; hermetically sealed Aldis gunsights were standard equipment in British fighters by mid-1916; metal linked am-

---

* Lansdowne would die nearly six years later while commanding the USS *Shenandoah*.

† As a B-24 pilot, Jimmy Stewart went into combat himself in December 1943 with the 703 Bombardment Squadron. In parallel with his movie career, he remained in the USAF Reserves and eventually reached the rank of brigadier general.

munition belts replaced canvas types that expanded when wet and often jammed the gun; and magnesium or phosphorus was added to a round's hollow base that, when ignited, left a visible trail and produced a "tracer" by which pilots could correct their aim.

Through combat necessity, engine technology had rapidly advanced to the point where there was now excess thrust, and true acceleration was a reality. This allowed comparatively high rates of climb and increased a plane's turning ability, which made dogfighting possible and opened the door to the development and weaponizing of aircraft. The puny 12-horsepower Balzer-Wright engine of 1903 had given way to the Benz Bz.IIIb and the Hispano-Suiza 8BA, each producing 195 to 220 horsepower, respectively. Top speeds of single-seat fighters like the SPAD S.XIII were up around 130 miles per hour, an unimaginable speed just fifteen years earlier.

Most early engines were the *rotary* type; that is, the entire engine and the propeller spins around the crankshaft, which generated very little vibration and provided an extremely stable gun platform. Rotary engines are air-cooled, and much lighter than their liquid-cooled counterparts, so they weighed less, thereby producing more excess thrust for maneuvering. But as they are spinning about in the airstream, rotaries generate drag—and a lot of it. This was a problem in the quest for higher performance, since gaining more power meant adding additional cylinders, or increasing the size of those available. Bigger cylinders would displace more pressurized air for combustion, but such an increase in size also drastically increased the engine's frontal area, and therefore the drag. This also equated to a higher fuel consumption, sometimes 25 to 30 percent more, above other types of engines. Given these limitations, the maximum available from a rotary engine was about 300 horsepower.

To overcome this limitation, the development of *stationary* engines that remain fixed while the crankshaft spins took precedence after 1916. With this arrangement, more cylinders can be added in various configurations rather than merely making bigger cylinders. Inline designs placed them along the crankshaft, while a radial engine arranged the cylinders in a star shape. The 1918 Liberty was an outstanding example of a "V" configuration where cylinders were angled up and away from the shaft. Weighing in at 845 pounds (dry), the obvious drawback was weight, as stationary engines were liquid cooled, and a comparable rotary engine like the German Oberursel UR.II would weigh about 150 pounds. But the trade-off in power was well worth it; a Liberty produced nearly 450 horsepower against 135 horses from the Oberursel.

But even the best engine possible is still dependent on two crucial components: fuel and the propeller. Piston engines produce power from the internal combustion of fuel and air that is metered by a carburetor, injected into the cylinder, and compressed by a piston. This "packed" fuel is then ignited by a spark plug; it explodes, and the resulting exhaust drives the piston up and down. This linear motion is converted to a spinning motion, either by the engine itself or by its connection to a crankshaft, and this drives the propeller. One problem has always been maximizing the efficiency of the engine; that is, compressing and converting every bit of available fuel to generate the most power. It was discovered that by adding chemicals, beginning with lead, fuel could be compressed further before combustion, and this greater compression resulted in more powerful explosions.* This produced more available thrust, which, all things being equal, gave a fighter greater potential maneuverability and better options in combat.

Yet improvements in fuel and engines would be obviated without parallel advances in the propeller. The tip of the spear, as it were, the prop converts all the energy produced by the engine into the forward motion that creates airflow over the wings, which, in turn, produces the lift required to fly. Though understood to be an airfoil itself, refinements in propeller design tended to lag, and it became quickly apparent that at about 1,500 revolutions the engine was operating faster than the prop could spin. Reduction gears, which transmitted the engine's energy but not its speed, were the answer and entered widespread use after the war.

With the transition of the airplane into a weapon came a corresponding requirement for greater control to maximize its maneuverability. For a fighter, control was critical because without the capability to accurately employ weapons, the whole aircraft was simply an aerobatic machine. Larger rudders were designed, as were elevators and wing flaps, though the latter were used sporadically during the Great War. Wing warping, which the Wrights had patented and jealously guarded, was definitely now of marginal utility due to the increased speeds available. With warping the pilot had to physically manipulate

---

* Always in danger from naval blockades, Germany began researching synthetic replacements for gasoline during the Great War. The Bergius process permitted the conversion of coal (one resource which Germany possessed in abundance) to synthetic oil. This trend of innovative and advanced research would continue through World War II.

the wing to turn the aircraft, and there is a limit to human strength. All early forms of control depended on the pilot's muscles, but twisting/warping a wingtip would not work in aerial combat and was a much less effective type of lateral control than the "little wing," or aileron.

This was a movable, rectangular surface flush-mounted near the tips and aligned with a wing's trailing edge. It was hinged and could be operated from the cockpit via cables attached to a yoke, or control column. When the ailerons are moved, airflow over the wing is disrupted and a lift imbalance created. As one aileron raises, the pressure over that wingtip dissipates so lift decreases, and that wingtip naturally drops. Simultaneously, the other aileron deflects downward, which creates higher pressure under that wingtip; therefore, lift increases and that wing rises.

British scientist Matthew Piers Watt Boulton is credited with the first patent (No. 392) for a workable aileron in 1868. Grandson of Matthew Boulton, who, with James Watt, manufactured the hundreds of Boulton & Watt steam engines that industrialized England in the late eighteenth century, the younger Bolton was an amiable recluse by nature and quite wealthy.* He had no desire for public acclaim, and his revolutionary innovation was ignored to the point where it could be "discovered" thirty-six years later by French engineer Robert Esnault-Pelterie. The Frenchman considered the wing warping dangerous and reinvented the aileron for use on his 1904 glider, which flew

---

* The younger Boulton's middle name is a tribute to James Watt. Seventy-two years into the future, the Boulton-Paul Defiant fighter, nicknamed the "Daffy," would see action during the Second World War.

successfully. He also patented the control column, or joystick, which provided a simple, single control point much less cumbersome than a wheel or yoke. However, Esnault-Pelterie was not the first to envision tandem control as Wilbur Wright had used a movable rudder in conjunction with his wing-warping system. However, the Frenchman's beautiful little R.E.P 1 monoplane was the first to employ a control stick when it took to the air on October 10, 1907.

French aviation pioneer Henri Farman had seen a Wright flying demonstration in 1908 and also came away convinced that ailerons were much better than wing warping for lateral control. His 1909 Farman III was the first powered, manned aircraft to use trailing-edge ailerons in the modern sense. An ungainly dragonfly of an aircraft, the Farman had a forward elevator with the propeller mounted "pusher" fashion behind the pilot. Though Wilbur and Orville Wright had unquestionably been first to succeed in manned, powered, and controlled flight, their real gift lay in adapting and/or improving existing technology: Lilienthal's airfoil, Langley's internal combustion engine, and Cayley's conception of a biplane with vertical and horizontal control surfaces. It was in this last area that the Wrights absolutely shone: control. They were the first to successfully design and implement, albeit awkwardly, a three-axis control system that permitted true flight.

Nevertheless, it was in Europe that aviation technology surged ahead for the next fifteen years, and blame for that stems from two primary sources: the Great War and with the Wrights themselves. The brothers were obsessed with secrecy; they did not even inform the press of their successful 1903 flight, and the first eyewitness mention in print was an obscure article in an equally obscure 1905 publication of Amos Root's

Gleanings in Bee Culture. For this reason the French, and specifically Alberto Santos-Dumont, believed Europe conquered the air before America. Santos-Dumont was the son of a wealthy Brazilian coffee planter who spent most of his adult life in France free to pursue his passion: aviation. In 1901 he circled the Eiffel Tower in a hydrogen-filled airship powered by a four-cylinder Buchet engine, winning the 100,000-franc Deutsch de la Meurthe prize.* Then, on October 23, 1906, he lifted off from the Château de Bagatelle in the Bois de Boulogne under his own power, holding a sixteen-foot altitude for a lateral distance of 197 feet.

Most of the world, except the few witnesses to the Wrights' 1903 flight, was certain the honor of man's first powered, controlled airplane flight belonged to Santos-Dumont and to France. Even after the Brazilian set the first Federation Aéronautique Internationale world record in November 1906, the Wrights still dismissed him as a fraud.† It didn't help that the Wrights had hangared their aircraft for three years, refusing to display it for fear the design would be stolen before they could sell it to the U.S. War Department. They also tried to license their craft for $25,000 a copy to France, Britain, and Germany but had little success except with the French, who, after recognizing that the brothers had indeed succeeded in 1903, initially agreed to buy manufacturing rights.

Other than greed, perhaps the most disconcerting trait the Wrights displayed was their refusal to recognize that others

---

* Santos-Dumont donated half the proceeds to his crew and gave the other 50,000 francs to the poor.

† He flew 722 feet in 21.5 seconds on November 12, 1906.

might also be capable of innovations better than their own, and they obstinately insisted that their wing-warping patent covered *all* types of lateral control. The brothers expected a royalty on every aircraft produced; in effect, they felt entitled to a monopoly on aviation, and their litigiousness hurt North American aviation development considerably. As Wilbur wrote, "It is our view that morally the world owes its almost universal system of lateral control entirely to us. It is also our opinion that legally it owes it to us."

Such an attitude was hardly conducive to the free exchange of ideas, nor was it good business. Alberto Santos-Dumont purposely never patented his own designs as he believed aviation would cultivate closer international relations and promote world peace. The brothers *could* have used their head start to become the elder statesmen of aviation—at a reasonable price for their efforts—but they did not. They also eschewed public flying demonstrations and aviation competitions because, as they stated in 1906, "We would have to expose our machine more or less, and that might interfere with the sale of our secrets." Fanatically litigious, Orville bluntly stated that "We did not intend to give permission to use the patented features of our machines for exhibitions, or in a commercial way."

Nor did they.

Some designers like Glenn Curtiss simply continued independent work and forced the Wrights to keep taking him back to court. In the summer of 1908 Curtiss sold an aircraft for $5,000, one-fifth of the price for a Wright aircraft, to the newly formed Aerial Experiment Association (AEA). This aircraft, named the June Bug, won the 1908 Scientific American Trophy, worth $2,500, and featured triangular ailerons on the wingtips.* These were connected to a shoulder yoke worn by the pilot so when he leaned in the direction of turn, the wires running from the yoke would move the ailerons. The Wrights were positively apoplectic. Lateral control in any form, they insisted, was their sole proprietary invention.

Glenn Curtiss maintained that movable surfaces on the back of a wing that altered its aerodynamic qualities was not wing warping, but a wholly new type of control. In 1909 a long, bitter court fight ensued and, in the meantime, Curtiss went right on improving and marketing his aircraft. Recognizing the value of public relations and competitions, he was quick to win the 1909 Bennett Trophy with a 47 mph speed record. During the 1909 Rheims Air Meet, every one of the Wrights' altitude and speed records was broken, and Louis Blériot flew his Type XI monoplane from Calais to Dover in thirty-six minutes, thirty seconds.†

Curtiss took note of all this. He had a knack for cherry-picking good ideas and discarding the bad, then ingeniously incorporating these improvements into his own designs. As a result, he quickly surpassed the Wrights, and by 1914 the Curtiss Aeroplane Company was the largest manufacturer of aircraft in the United States. In the end, the brothers spent so much time fighting in court that they were left behind in the very field they had pioneered. Avarice, it seemed, had triumphed over aerodynamics and innovation. As Wilbur himself

---

* Wilbur declined to compete against Curtiss, whom he loathed, and was convinced his ideas would be stolen.

† Blériot utilized the control stick invented by his countryman Robert Esnault-Peltrie, rather than the cumbersome multilevered Wright system.

rather blatantly confessed, "I want the business built up so as to get the greatest amount of money with as little work."

Meanwhile the AEA, under the leadership of Dr. Alexander Graham Bell himself, set out to advance aviation interest and development through a "cooperative association" between like-minded men and ideas. By virtue of his status as a flyer, businessman, and inventor, Curtiss was also asked to join and Augustus Post, one of America's least-known but more-intriguing figures, served as a representative from the Automobile Club of America. Scion of a wealthy banking family, Post could best be described as a gentleman adventurer and bon vivant. A graduate of Amherst and Harvard, he was a talented bass and sang with the New York Symphony Chorus. Also an avid balloon and auto enthusiast, he owned the first automobile in New York City, built the original parking garage, and was the founder of the American Automobile Association (AAA).[*]

Post saw the tremendous potential of aviation and penned a 1914 article that stated, "A man is now living who will be the first human being to cross the Atlantic Ocean through the air. He will cross while he is still a young man. All at once, Europe will move two days nearer; instead of five days away." He was quite correct; Charles Lindbergh was twelve years old when the article appeared in the *North American Review*. In fact, it was Post who suggested to Raymond Orteig that $25,000 be offered for the first aviator to cross between New York and Paris. The hotelier agreed and publicized the Orteig Prize in 1919:

---

[*] The garage was located at Sixty-Sixth Street and Columbus Avenue beneath the St. Nicholas ice skating rink.

Post wrote the rules.* A true visionary, Augustus was a fierce advocate for the science of aviation and believed in the importance of laboratories and experimentation. He would be among the first to realize the importance of specially designed aviation fuel that would increase performance and make way for powerful engines like the jet. Post could also see a time when rockets would send men beyond Earth's atmosphere, and he described a futuristic Interplanetary Society very much like NASA.†

So while the Wrights sued everyone with competing interests, the Europeans rapidly advanced through the assistance of government-sponsored organizations like the British Advisory Committee for Aeronautics and the Royal Aircraft Factory at Farnborough. In 1904, renowned German engineer Ludwig Prandtl published "Fluid Flow in Very Little Friction," which accurately described the concept of boundary layer separation and the consequences of it. France had long been at the forefront of aviation, and since the late nineteenth century aviators of all sorts conducted various experiments at Chalais-Meudon, just north of the city.

By 1906, as Gustav Eiffel concluded his air resistance experiments from the second level of the tower bearing his name, the French Central Establishment for Military Aeronautics had been created. Clearly the Europeans realized the significance, both civil and military, of the aircraft. "In an age of intense nationalism," Michael Gorn perceptively writes in *Expanding*

---

* The very prize that Charles Lindbergh would win by crossing nonstop from New York to Paris in 1927.

† Post also published numerous magazine articles, authored a book titled *Skycraft*, and starred in an opera: *The Man from Paris*.

the Envelope, "on a continent where states lay in close proximity, every advanced government sought to guide and nurture this powerful but unknown technology."

Meanwhile, due to infighting, politics, and general official disinterest, the development of American aviation lagged considerably. Finally, after years of failures and finally awake to the looming danger of war in Europe, the Smithsonian Board of Regents reopened Langley's laboratory. In early 1915 the Main Committee of the Advisory Committee for Aeronautics was formed to coordinate, oversee, and generally consolidate the efforts of several diverse aviation departments. By spring it had changed its name to the National Advisory Committee for Aeronautics; the NACA was born, and with it came the possibility of a national research center to rival the Europeans.

Though often plagued by bureaucratic competition or official indifference, the new organization was appropriated $85,000 and immediately began making a difference. The nasty patent dispute between Wright-Martin and Curtiss Aeroplane was settled by the U.S. Circuit Court of Appeals, which incidentally upheld the Wrights' claim. Curtiss shrugged, paid the fine, continued what he was doing, and eventually acquired the Wright Company (later Wright-Martin) after Orville sold out in 1915.* The NACA hired top-notch professional engineers, technicians, and draftsmen; began detailed experiments with propeller design; and, perhaps most significantly, mediated an

---

* America's entry into World War I finally ended the whole mess when the government mandated cross-licensing of all patents relative to war production in order to streamline effort.

agreement between automobile manufacturers like Packard, Lincoln, Cadillac, and others to produce an aircraft engine.[*]

It was through this merging of available technology during the Great War that aviation shot forward into its next great era. There truly were no limits once engineering and aerodynamic knowledge caught up with man's ingenuity, and this ushered in a "Golden Age." It was during this time, from the final days of the Great War through the beginning of the Roaring Twenties, that the men who would relentlessly chase the demon were born. A volatile, unsettling time, the 1920s were also filled with promise, excitement, and imagination. It was, in this author's opinion, *the* formative decade of the last one hundred years and set in motion events that still shape our world today. A perfect storm of economic and societal causes incubated the German national socialism, Italian fascism, and Japanese imperialism that led to the Second World War. Europe's woes, and by default those of the rest of the world, and the failure to recognize the dire consequences those woes would inflict, can be traced to this decade. There were those in America who could see the war coming, but they were largely ignored.

The United States emerged from World War I like a teenage boy surprised by his newly discovered muscles. Though the industrial flexing of 1914 to 1918 was minuscule compared to what would come during the Second World War, it was enough to awaken the potential of the United States, and in loaning the Allies the money needed to purchase military matériel, America overtook Britain as the largest creditor nation on Earth. By

---

[*] The result was the Liberty L 12.

the early 1920s U.S. foreign investments totaled nearly $16 bil-
lion, with $4 billion to postwar Germany alone.

Herein lay the problem.

The fragile and insecure Weimar Republic had to make good
on 132 *billion* gold marks, or about $32 billion in U.S. dollars,
in war reparations to Belgium, France, and Great Britain for
Imperial Germany's "Criminal Pride."* In 1919 Georges Clem-
enceau, premier of France, said of the latent Teutonic threat,
"The next time, remember, the Germans will make no mistake. They will break through into Northern France and seize
the Channel ports as a base of operations against England."

Unfortunately for the world, he was prophetically correct,
and a large slice of blame lies with political decisions that fostered the bad economics. Subjectively, the Allied powers' thirst
for vengeance can be understood, but objectively it was a disaster and many diplomats disagreed with the terms, rightly fearing the future threat of a desperate Germany. The World War
Foreign Debt Commission was established in February 1922,
one month after Bob Hoover's birth, to manage the repayment.
Fifteen separate agreements were signed based on each nation's ability to pay, with terms extending sixty-two years into
the future.†

In 1924, when George Welch was six and Ken Chilstrom

---

* Through some very convulsed phrasing, bonds, and other measures, Germany was actually scheduled to pay about half this amount. Nevertheless, it
was unsustainable and payments were regularly defaulted until 1932 when
they ceased altogether.

† It still wasn't enough for several countries. Germany finally paid its debt
in October 2010, while Great Britain ceased making payments on its $4.4
billion Great War debt in 1934.

was a toddler, Washington loaned Berlin the funds needed to meets its postwar debt service. With Germany now fulfilling its obligation, then France, Belgium, and Great Britain could repay the $7 billion owed to the United States—with interest—and everyone would win. This would, it was hoped, permit Europe to recover and resume full trade, including significant imports from America and thereby bolstering both economies.

After a brief postwar recession, America recovered and had an unprecedented, though relatively short, burst of prosperity and growth. Parents such as George Lewis Schwartz or John and Emma Chilstrom would have been fascinated, as America was in general, with amazing new products flooding department stores. Toasters, vacuum cleaners, telephones, and refrigerators to name just a few; automobile production was surging and there were more cars in New York alone than in all of Germany. Dupont Chemicals, where George Welch's father worked as a senior research chemist, had patented a new fast-drying, hard lacquer and suddenly there was a rainbow of color choices for cars other than Ford's boring, utilitarian black.

By 1925 the radio craze had struck in full force and the WSM Barn Dance was on air in Bob Hoover's home city of Nashville.* Over $60M was spent on sets and spare parts, while the number of broadcasting stations increased from 30 in 1922, to over 500 by 1923. Radio shows like *Burns and Allen* or *Fibber McGee and Molly*, both of which premiered in the 1930s, kept listeners across the country amused, while NBC and CBS news programs brought world events into living rooms across the

---

* It became the Grand Ole Opry in 1927.

country. During July 1921, three months after Ken Chilstrom was born, 1 in 500 American homes had radios and listened as the Demspey-Carpentier fight changed sporting events into mass entertainment.

Kids growing up in the 1920ss would have been enthralled with the *House of Myths* or *Amos 'n' Andy*; Ken Chilstrom thrilled to *Uncle Don's Strange Adventures for Children* as his parents relaxed with jazz, opera, or show tunes from commercially sponsored shows like *The Fleischmann's Yeast Hour*, or even the *Champion Spark Plug Hour*. At least six million American radios were tuned to Graham McNamee as Charles Lindbergh arrived in Washington on June 11, 1927, and during the following summer came *Plane Crazy*, the film debut of Mickey and Minnie Mouse.

But there were problems, of course.

By 1920 approximately 30,909 national and state banks existed in the United States, and many were not the solid, secure institutions of the previous century. Created ad hoc to take advantage of the rapidly inflating economic bubble, many of these banks were run by less than qualified individuals who used them to play the decade's great game: speculation. Real estate was a favorite, especially in Florida and California. But a 150 mph storm hit Miami and Fort Lauderdale head-on in 1926, and at that time was called "the most destructive hurricane ever to strike the United States." Two years later, during the night of September 16, 1928, another one roared ashore, flooded Lake Okeechobee near Palm Beach, and the dikes dissolved. Several thousand were confirmed dead and property damage was significant.

California (unsurprisingly) simply overdid it. Real estate permit values increased tenfold in "value" over a few years, and by middecade a developer was quoted saying, "We have enough subdivisions and lots for sale and in process of development to accommodate the cities of New York, Philadelphia, and Detroit." With East Coast real estate awash in storm water and a bloated West Coast market, the banks panicked and began pulling in their notes.

Stocks were the other speculative El Dorado. The Great Bull Market of 1927 to 1928 saw 577 million and 920 million shares traded, respectively. The chance to get rich quick was intoxicating, and broker loans ($8.5 billion by September 1929) allowed people to invest with only 10 percent down while the brokerage house carried the rest on margin. Everyone, it seemed, was jumping aboard the prosperity train, including farmers who mortgaged their land and amateurs gambling with money they did not have.

It could not last.

Like tremors before an earthquake, all the signs of an impending financial calamity were in place by the summer of 1928. Tens of thousands of farmers had mortgaged their properties to play the market, yet with agricultural prices falling they began defaulting on their overvalued, heavily indebted farms, and at least 40 percent of the rural banks in business in 1920 had by early 1929. With Herbert Hoover's nomination in June 1929, optimism rebounded and the volatile situation stabilized—temporarily. In his June acceptance speech Herbert Hoover gave voice to the national optimism by stating, "We in America today are nearer to the final triumph over poverty than ever before in the history of any land. The poorhouse vanishing from among us . . . we shall soon, with the help of God, be in sight of the day when poverty will be banished from this nation."

Little did he know. But others did, and savvy investors like trader Michael Meehan and GM executive John Jakob Raskob discreetly sold off large chunks of stock while the professional bankers and businessmen continued to worry. The average investor didn't understand or care, as long as the brokerage houses continued to hold 300,000,000 shares on margin and the dividends kept rolling in. The market dipped in December, then, in February 1929, sharply collapsed. Margin calls forced sell orders and a record 8,246,740 shares changed hands that month. The Federal Reserve Board refused to act and publicly stated that it would not "contemplate the use of the resources of the Federal Reserve Banks for the creation or extension of speculative credit."

It was an understandable position but, in this case, short-sighted and disastrous. Bankers like Charles Mitchell of the National City Bank in New York well understood the peril, and his institution alone advanced $20 million to cover the calls, keep the brokerage houses above water, and the market alive in hopes that it would stabilize. Economists moaned but the general public ignored it. It was, the uninformed or greedy said, just a correction, and they kept buying in a frenzy of optimistic denial. Investment trusts, much like modern hedge funds, proliferated, and speculative ventures of all types permeated the American economy.

And why not? Skirts were shorter than ever before, and Prohibition was a dead letter; skyscrapers continued to climb, roads were improved, and automobiles were everywhere. Hemingway's *Farewell to Arms* was released along with William Faulkner's *The Sound and the Fury*, and Erich Maria Remarque's huge bestseller *All Quiet on the Western Front* hit the shelves. Even though Voltaire's *Candide* was banned in Boston, one could sway to the rolling hoarseness of Louis Armstrong's "Basin Street Blues," or drink illegal martinis to Ben Selvin crooning "My Sin." That year also found Commander Richard Evelyn Byrd shivering in Antarctic darkness waiting to fly to the South Pole, with the *New York Times* headline in late May exuberantly proclaiming:

COLONEL LINDBERGH WEDS ANNE MORROW IN
HER HOME; MAY FLY ON HONEYMOON

Aviation stocks remained among the highest, with Curtiss reaching nearly 150 points while Wright Aeronautical, now the premier manufacturer of engines and basking in the light of Lindbergh's successful flight, approached 300.* The Federal Reserve Board bumped its discount rate to cool the fever, but it had little effect as investors simply borrowed from other sources. Five months after Ken Chilstrom's ninth birthday the Dow Jones Industrial Average hit a record of 381.17 on September 3, but on Tuesday, October 24, the cracks widened and spread.

By Thursday everything from U.S. Steel to Montgomery Ward crashed. People panicked and, as before, bankers acted quickly to minimize the disaster. Their motives were practical, of course, not altruistic. They knew if they could restore some sense of balance and prevent a total market collapse, then measures might be taken to repair the damage before it became a catastrophe. This time six men, heads of the most important

---

* It was a magnificent Wright Whirlwind J-5C that took Lindbergh from New York to Paris in May 1927.

U.S. banks, met in the Wall Street offices of J. P. Morgan and Company across from the Exchange. In a matter of minutes they each pledged $40 million from their respective institutions, $240 million in all, to prevent an economic meltdown. The market, they believed, would recover if certain key securities were stabilized and margins met.

It worked.

Panic subsided and by the end of the day, after some twelve million shares changed hands, stocks actually closed higher in several cases. Friday and Saturday was still a roller coaster but on Monday, after the fear had passed, jittery investors decided to cash out while they could rather than risk another Black Thursday. So, on October 29, a massive dumping of stocks occurred and the result was an even blacker Tuesday, with 16,410,030 shares traded by the closing bell. This time it set off a chain reaction among the other North American exchanges, including San Francisco, Chicago, and Toronto, then the major foreign exchanges in London and Paris. Like spilled paint spreading across a floor, the consequences spread and amplified, exacerbated by diverse causes but financially apocalyptic in effect. Fortunes were lost and made. Society was irrevocably altered as men and women adapted to their new reality. Though some men deserted their families in the face of fear and uncertainty, most did not. The Chilstroms, Welchs, and countless others like them took a deep breath, mastered the dry-mouthed fear that comes from fearing for a family, and persevered.

They held on.

Though their faith in government paternalism, banking security, and the wisdom of businessmen was forever damaged, they did not lose faith in themselves or the hope represented by their children. This mental agility, perseverance, and *toughness* to endure a disaster none had foreseen was passed on to their children. Teenagers like Ken Chilstrom and George Welch, who would go on to force the world back into a livable place and, in so doing, also force the demon out into the open.

Part Two

===

INTO THE FIRE

There are no great men, just great challenges
which ordinary men, out of necessity, are
forced by circumstances to meet.

FLEET ADMIRAL WILLIAM FREDERICK HALSEY

Three

=========

The Next Leap

By the mid-1930s America was recovering from what Lionel Robbins, a British economist, called the Great Depression. Though he was arguably the first to capitalize the phrase, president James Monroe used the same words in his 1820 Fourth Annual Message, and the 1928 Republican Party proudly stated, "Under this Administration [Coolidge] the country has been lifted from the depths of a great depression to a level of prosperity."

Still, though the Big Bull Market died and took with it literally billions in profits, this did not, by itself, provoke the Great Depression. The president reacted relatively decisively with a public works program to stave off unemployment, and tax cuts to stimulate the deflated economy so, by the New Year, there was a tentative recovery in progress. A recovery that might have survived, lessened the financial impact here and abroad, and very possibly prevented, or at least delayed, the international malaise leading to the Second World War.

Most critical, and often the least discussed, was the failure of the U.S. Federal Reserve system to do exactly what it was created to do: prevent a financial catastrophe. Twelve member banks with their respective board of governors were backed by the government and responsible for several key functions. They would set the discount rate, the prime interest rate charged by Federal Reserve banks to loan money to commercial banks. By regulating this, the economy could, in theory, be manipulated to either slow or stimulate growth as conditions dictated. The Fed also issued paper banknotes and determined the reserve limits of cash each member bank kept on hand. It was from this reserve that monies could have been paid out to smaller institutions to prevent the 1,352 bank failures in 1930 alone. Finally, in late December 1930, the Bronx branch of the Bank of the United States was forced to close its doors after losing $2 million in panicked withdrawals. This caused a wider panic as word spread that banks were closing and by the end of the year over $550 million in deposits had been liquidated.

But the Federal Reserve did nothing.

Then, the month following George Welch's twelfth birthday, the U.S. Congress enacted the Smoot-Hawley Tariff Act. A protectionist piece of legislation, this was theoretically authored to protect American agriculture from foreign imports and pushed through by President Hoover in fulfillment of a campaign promise, a promise made before the crash of 1929 and signed into reality without consideration of its effects on the weakened economy. More than this, the act triggered retribution from European markets that decimated foreign trade on both sides of the Atlantic, pouring fuel on simmering nationalistic fires in Germany, Italy, and Japan. During the September Reichstag elections, the National Socialists, or "Nazis" for short, won over 18 percent of the vote, becoming the second-largest party in Germany.

Then came the drought in North America.

It began in the dry summer of 1930 and became a full-blown disaster in 1934, 1936, and 1939. Sparse rainfall turned over 100 million acres of the Oklahoma and Texas panhandles into a barren wasteland, which later expanded into Kansas, Nebraska, Colorado, and New Mexico. Winds from within the "Dust Bowl," as it was subsequently named, stirred up huge clouds called "black rollers" that blew as far as the East Coast. Farms disappeared and a mass migration of homeless, destitute Americans trudged out of the Bowl looking for a fresh start and exacerbating the overcrowding, unemployment, and desperation in large urban centers.

It was into this world then, that the adolescent demon hunters were growing to maturity. Ken Chilstrom's father had lost his job at J. C. Penney's, and like thousands of others had to migrate to the nearest big city in hopes of finding work. In this case it was Chicago, and the little suburb of Elmhurst. Though not inundated by mass communication or instant news, Ken was certainly aware of some issues facing the country, at least as it affected him. He did realize that all was not right in his parents' world, and, though too young to recall much of the Roaring Twenties Ken, like most children, were affected in varying degrees, mentally and psychologically, by the Great Depression.

"Money was scarce," Chilstrom recalls. "We didn't even have a radio anymore. I worked five jobs, including paper routes for the *Chicago Tribune* and *Liberty Magazine*."

Chuck Yeager, growing up in Myra, West Virginia, might have been the least affected, as he later wrote, "it had no real impact when you were already so low on the income scale." For him, shooting squirrels and rabbits for his mother to cook was just part of life. Through it all the children, many now teenagers like George Welch and Ken Chilstrom, absorbed their parents' resiliency to become self-sufficient, adaptable, and tough.

Yet in this chaotic, unsettling, and often desperate time, aviation provided, as it always had, inspiration and hope. On April 20, 1930—Ken Chilstrom's ninth birthday—Charles Lindbergh and his new wife flew from Los Angeles to New York in fourteen hours and forty-five minutes; a new coast-to-coast record. "I told my father I was enthralled with the idea of flying from hearing about Lucky Lindy. I think it was the first time I realized that this was what I wanted to do."

He was not alone. Thousands of little boys, and girls, felt the same way. Two weeks before Smoot-Hawley was signed, America was further awed when Lieutenant Apollo Soucek took off from Naval Air Station Anacostia and set a world altitude record of 43,166 feet in an open-cockpit Wright Apache.[*] By the end of the year Glenn Curtiss had died in Buffalo, New York, and Clarence Birdseye patented his quick freezing process. The spring of 1931 saw "The Star-Spangled Banner" officially adopted as the national anthem and shortly before George Welch turned thirteen, the Empire State Building opened to the public. Most dramatic, on June 23 Wiley Post and Harold Gatty lifted off their Lockheed Vega, *Winnie Mae*, from Roosevelt Field to circumnavigate the world in eight days.[†] This

---

[*] Soucek would go on to fight in World War II and was the executive officer of the USS *Hornet* when it sunk in 1942. He died in 1955 and NAS Oceana was dedicated in his honor.

[†] Eight days, fifteen hours, and fifty-one minutes.

was the same Long Island airfield where Charles Lindbergh left for Paris four years earlier, and the famous pilot was tragically back in the news himself in March 1932 when his baby, Charles Augustus Lindbergh Jr., was kidnapped. The toddler's little body was later found within miles of his Hopewell, New Jersey, home.

Nineteen thirty-two had some happier news, though, and there were glimpses of financial recovery, though the government numbers were so bad that no official statistical abstract was prepared for the year. On April 19, Robert Goddard launched a rocket that remained stable throughout its short flight by using gyroscopically controlled vanes, and in the fall Franklin Delano Roosevelt was elected the thirty-second president of the United States. Inaugurated on March 4, 1933, Roosevelt eloquently calmed the nation by stating, "This great nation will endure as it has endured, will revive and will prosper. So, first of all, let me assert my firm belief that the only thing we have to fear is fear itself."

Facing 25 percent unemployment and a disillusioned, largely disoriented public, FDR had a steep hill to climb and began immediately. Within a week of his taking office, the New Deal, a series of initiatives designed to break the country out of the Great Depression, was enacted. One of these, the 1933 Banking Act, established the Federal Deposit Insurance Corporation, which guaranteed $2,500 worth of deposits for each account holder. This, plus the Federal Reserve's issuing of more money, began to instill confidence once again in the nation's banks. A young Chuck Yeager, along with three million others, was eventually enrolled in the Civilian Conservation Corps, and work projects sprang up all over the country. The CCC took young men, 70 percent of whom were malnourished, with no

skills and limited education, and gave them food, shelter, clothing, and medical care. In return for forty hours of work per
week they were also paid $30, most of which had to be sent
home to their families. These young men signed up for a minimum of six months and built roads and bridges, planted some
three billion trees in national parks, and worked in a variety of
infrastructure projects.*

It was a start.

With the banking system shored up by the Federal Reserve
and millions of men back to work making noticeable improvements, confidence began to return. In July 1933 Wiley Post astounded the world again by flying solo around the globe, taking
off from Floyd Bennett Field at the mouth of Long Island's Ja-
maica Bay. Post returned to a crowd of 50,000 people gathered
at the same airfield seven days, eighteen hours, and forty-nine
minutes later: twenty-one hours faster than his previous time.
Two months later, over Villacoublay, France, Gustave Lemoine
reached 44,819 feet in his open-cockpit Potez 506.

Roscoe Turner, daredevil and adventurer, set an American
coast-to-coast speed record in his stubby little Wedell-Williams
racer of ten hours, four minutes, and thirty seconds. A young
Bob Hoover had seen Turner perform at Berry Field outside
Nashville and recalled that the pilot was "the closest thing to a
hero I had ever seen." He certainly looked the part, always perfectly turned out in a tailored British officer's uniform complete
with riding pants, boots, and a long white scarf. Turner sported

---

* Other enrollees included actors Robert Mitchum, Raymond Burr, and
Walter Matthau; future U.S. representative Ed Roybal, and baseball legend
Stan Musial.

an immense, waxed handlebar mustache and toured with a magnificent lion he'd named Gilmore, in honor of the Gilmore Oil Company, his sponsor. Ken Chilstrom agreed. "I'd seen him fly and got to speak with him afterwards. I was awed. I wanted to look like that. I thought that's how all pilots should look!"

Yet as the clouds lifted a bit in America, in other places they noticeably thickened. In Germany the 1933 Enabling Act was passed, which permitted the chancellor, Adolf Hitler, to enact laws as he saw fit without using the Reichstag—in effect transforming the Berlin government into a dictatorship. Though Germany ominously pulled out of the League of Nations in October, the year ended well, at least in the United States, as the National Prohibition Act died a welcome and long-overdue death. Roosevelt, keenly aware of the prevailing political wind, knew no other amendment had been repealed in America's 140-year constitutional history, but he also knew this could be another welcome break with the past, and a further means of improving national morale.

Shortly after his inauguration the president managed to legalize the sale of beer by redefining the meaning of intoxication, so the Methodist Board of Temperance, Prohibition, and Public Morals, and others of that ilk, were fighting a losing battle. By midsummer fourteen of thirty-six state legislatures had ratified the Twenty-First Amendment repealing Prohibition, with Utah signing last on December 5, 1933. "Booze cruises" were over; speakeasies faded into history, and much of the illicit pleasure in drinking vanished overnight, yet Americans generally rejoiced.

So did Washington, in fact.

Besides the reenactment of a slew of federal crime laws, nearly $260 million in alcohol taxes were collected in the year following Prohibition's demise. So large was this, 9 percent of the federal budget, that Roosevelt gave taxpayers a sweeping income tax cut that not only stimulated the economy, but also boosted his popularity to the point where he was politically un-stoppable, an attribute that would pay huge dividends as war clouds once again gathered over Europe. That, and the enor-mous influx of gold to the United States from European coun-tries who could read the writing on the wall.

There were portents of this all by the mid-1930s, after Ger-many's Night of the Long Knives and the creation of the Dachau concentration camp for political prisoners. Two laws were quickly passed in Nuremberg that stripped Jews of their Ger-man citizenship and made intermarriage illegal. There were also increasing signs that aviation would play a significant role in whatever happened next. Deutsche Lufthansa, the German state airline, commenced the first transoceanic airmail service between Stuttgart and Buenos Aires while a Soviet balloon, the *Osoaviakhim*, soared thirteen miles into the stratosphere. Dr. Rudolph Kühnold detected an aircraft on his primitive radar set and George Welch, halfway through St. Andrew's in Dela-ware, was considering a career in engineering. John Chilstrom had sufficiently recovered financially to open a grocery store, "Chilstrom & Burke," in Elmhurst, and young Ken built a gas-powered aircraft with a five-foot wingspan.

Over the next few years most of the economy, and the pros-pects of young Americans, improved dramatically. With blunt financial force, the federal government continued muscling the country from the grip of depression. Bob Hoover, now a teenager, was fascinated with aviation and read voraciously on the subject, picking Jimmy Doolittle as his hero. "He was my true idol," Hoover later wrote in *Forever Flying*. "I wanted to be just like him . . . the only thing I ever wanted to do was fly airplanes." Ken Chilstrom voiced similar sentiments. "He [Doolittle] was a very outgoing guy. A real person. A rare person."

Doolittle had been a flight instructor during the Great War and could fly anything with wings. Just as vital, he understood the technical details and engineering aspects as few pilots or engineers did. He recalled, "In the early '20s, there was not complete support between the flyers and the engineers. The pilots thought the engineers were a group of people who zipped slide rules back and forth, came out with erroneous results and bad aircraft; and the engineers thought the pilots were crazy."

But Doolittle was no basic stick-and-rudder pilot. He had graduated from the University of California, Berkeley, then went on to get a master's of science and a doctorate in aeronautics from the Massachusetts Institute of Technology. No deskbound academic, Doolittle won the Schneider Cup and Mackay Trophy in 1925 and 1926, respectively. He was the first pilot to successfully fly an outside loop, and the first to make a complete flight, from takeoff to landing, on instruments. Like Lindbergh, Doolittle recognized that aviation had to be freed from weather constraints if it were ever to progress. Resigning his regular army commission in 1930, he went to work for Shell Oil, lending expertise to, among other things, the development of 100 octane fuel. Continuing to fly, Doolittle won the first Bendix Trophy in 1931.

Through it all the nation marched on and war seemed increasingly imminent, at least in Europe and Asia. Germany formed the Luftwaffe in 1935, and the following year occupied the Rhineland. German physicist Hans Joachim Pabst von Ohain had studied under Ludwig Prandtl and earned a PhD in physics and aerodynamics from the University of Göttingen.

Von Ohain, who designed the first operational turbojet engine, had independently arrived at many of the same conclusions regarding gas turbines as had a young pilot named Frank Whittle. The German was as unaware of Whittle as the Englishman was of von Ohain. Interestingly, in 1928 the young British pilot had authored "Future Developments in Aircraft Design," a thesis he had written while a flight cadet at RAF Cranwell. The astounding aspect of Whittle's work was that the twenty-one-year-old pilot described what he called a motorjet, essentially a piston engine where compressed air was funneled into a combustion chamber.* "I was thinking in terms of a speed of 500 mph in the stratosphere at heights where the air density was less than one quarter of its seal-level value," he wrote.

After graduating from Cranwell, Whittle was briefly posted to No. 111 Fighter Squadron at Hornchurch, then to Wittering in the Flight Instructor's course. Like Jimmy Doolittle, the young officer possessed the invaluable combination of intuitiveness and formal education, which, when harnessed with considerable flying skills, produced the best type of test. His work initially centered around improving existing technology, utilizing a conventional reciprocating engine to power low-pressure fans rather than propellers by expelling the heated exhaust through a nozzle. But due to the pressure ratios available with internal combustion there was only so much compression available, and Whittle *knew* there was a better method, something so revolutionary that current speed and altitude restrictions would be meaningless if only he could find it.

---

* At Cranwell, Whittle also authored *Sea Power in the Pacific*, where he predicted a Japanese surprise attack on Pearl Harbor.

While assigned to Wittering, his thinking led Whittle beyond conventional solutions and into new territory. It was here, in 1929 and 1930, where he dispensed with the heavy, fuel-eating reciprocating engine altogether and envisioned heated air-driving turbine blades that subsequently powered a compressor: the essence of a turbojet engine. To achieve the power, and therefore flight past 500 mph and above 40,000 feet that he foresaw, Whittle knew air would have to be compressed to a much greater degree than that possible with conventional engines. The simplest way to do this was to increase the velocity of the incoming air, which would subsequently raise its pressure.

But how?

The solution, which astonished Whittle, was simplicity itself: in theory. By 1930 there were significant technical and metallurgical hurdles to overcome before an operational, sustainable jet engine could be fielded, but the principle for his engine—for all jet engines—is essentially the same. Air is sucked into a series of bladed fans (compressors) that spin and accelerate it, thus raising the pressure. Each fan is a called a "stage," and as it passes through each stage the airflow velocity increases. As this occurs the pressure also rises, resulting in a stream of very fast, pressurized air that is forced into a combustor. Smaller than the intake and compressor, this chamber accelerates the air, like a wide river that suddenly narrows, as it bursts into the combustion section. Nozzles spray fuel into the high-pressure air and the mix is ignited.

The resulting explosion produces extremely hot gas that expands and must escape. The only way out of the combustor is through the exhaust section, or nozzle. It is this hot, high-pressure exhaust blasting backward that thrusts the jet forward. In Frank Whittle's initial design the gas would turn the shaft, spin the compressors, and expel sufficient exhaust to produce thrust, yet it was impractical because it depended on a conventional piston engine to draw in air for compression. His revolutionary solution, and one that differed from von Ohain and other parallel attempts, was adding together several stages of multiple-bladed fans to create a turbine. The high-pressure exhaust gases spin the turbine that is connected by a shaft to the compressor. By rotating, the turbine sucks in more air to be compressed, combusted, and expelled and will continue doing so as long as there is fuel. The resulting self-contained engine is called a turbojet.

By the time the market crashed in October 1929, Frank Whittle had refined his original thesis into a largely practical propulsion system. But if he was awed and thrilled by his idea, the scientific and military communities were less excited. William Lang Tweedie from the Directorate of Engine Development was painfully direct. He informed Whittle that the Air Ministry did not look favorably upon gas turbines and cited a 1920 Aeronautical Research Committee Report (No. 54) that damned the entire notion.

Dr. Alan Arnold Griffith of the Air Ministry's South Kensington Laboratory also shot Whittle's design down in short order. Griffith, a renowned mechanical engineer and expert on metal fatigue, opined that the young pilot's optimism had colored his mathematical calculations; the engine was far too heavy and would produce insufficient thrust to be of any operational use. In Griffith's case his motivations were fairly obvious; he had been seeking Air Ministry support for his own design, which utilized the gas turbine to power a propeller. To his credit, he overcame a major design issue with existing turbine blades by realizing they were, in fact, small airfoils and needed to be constructed that way. However, to Griffith's lasting discredit, he would not advocate Whittle's lighter, cheaper, and better design out of avarice and, quite possibly, professional jealousy. The general conclusion from these prominent experts, and a conclusion that affected government funding, was that gas turbines were too heavy and the power they produced was not enough to justify the extra weight.

At the time, they were quite correct.

The gas turbine was powered by an internal combustion engine that was employed to spin a propeller. This was not what Frank Whittle proposed and, discouraged but undaunted, he filed a Provisional Specification for his turbojet on January 16, 1930.* But life, as often happens, was catching up with him and he married his sweetheart, Dorothy May Lee, in May. A son was born the following year, and Whittle began test work on the Royal Navy's Fairey III aircraft. This necessitated a carrier takeoff and landing qualification, so by July 1932 Whittle had logged seventy-one catapult shots, completed the testing, and been posted to the Officers Engineering Course at RAF Henslow.

During 1934 the twenty-seven-year-old flight lieutenant was selected to attend Cambridge University and he graduated in 1936, with honors, in mechanical sciences. Immersed in academia, busy with flying and his family, Whittle's interest in turbojets faded until he received a letter from an old RAF pilot friend who indicated he had discovered a source of private fi-

---

* British patent No. 374,206.

nancing for the development of the jet engine.* As a serving officer, Whittle would normally have been excluded from engaging in private business but the Air Ministry, perhaps with an eye on the future, permitted a joint venture, and by mid-1936 his company, Power Jets Limited, was a reality.

Also a reality was increasing volatility from across the Channel and across the world. Adolf Hitler had become chancellor in 1933 and commenced Germany's rearmament immediately. Britain, at least the Royal Air Force, had noticed, and with the official creation of the Luftwaffe in February 1935, the Air Ministry redoubled its commitment to modernizing the air force. New monoplane fighters with variable pitch props, enclosed cockpits, and retractable landing gear had been designed and fielded. The Hawker Hurricane first flew in late 1935, with the Supermarine Spitfire following by March 1936—the same time Power Jets was formed.

In Japan an uprising of military officers had assassinated two former prime ministers and gained brief control over the Ministry of War. Though the insurgency failed, it gave the military an excuse to put active-duty officers in key cabinet posts, thereby giving the military a de facto veto power over the civil government. This permitted unchecked, aggressive expansionism in China, and ultimately full-scale war in the Pacific.

At the same time the Germans had a breakthrough of their own with jet technology and, realizing the jet's enormous po-

---

* The initial patent expired in 1935, and Whittle could not afford the £5 renewal fee so it lapsed. The investment banking firm of Falk & Partners provided the backing, and one of the directors was Sir Maurice Bonham-Carter, grandfather of the actress Helena Bonham-Carter.

tential, made rapid progress aerodynamically. While obstacles were overcome, engine technology remained a huge limitation. In Germany's case this was exacerbated by a shortage of precious metals, especially chromium, nickel, and titanium. These were essential for the production of Tinidur, an alloy utilized for jet turbine blades. More damning was the effect on engine reliability; a Jumo 004 had a life expectancy of 25 to 30 hours compared to 125 hours for Whittle's W.2/700 turbojet. Even at the engine's best, metallurgical problems were a constant issue, which was puzzling in a nation famous for its engineers.

Nevertheless, technology begets technology and the quest for speed—to be the fastest—was tantalizing in its murkiness. What would be possible if man could fly past the speed of sound? Did a barrier even exist and, if so, could it be breached? Then what? These were tantalizing thoughts for scientists, engineers, and military men. But such thoughts had to be shelved until it was proven that man and machine could survive. Until it was proven that the demon could be tamed.

Seven years after Whittle patented his jet, Hans von Ohain constructed a very simple, but workable, turbojet of his own, the HeS 1, at Ernst Heinkel's factory on Marienehe Airfield. Encouraged by its success, he immediately began improvements for a flight-capable jet engine. Faced with official indifference from the Reich Air Ministry, Heinkel funded the research and development as a private venture. The result was the Heinkel 178, a pretty, all-metal jet with high-mounted, elliptical wings and gear that retracted into the fuselage. On August 24, 1939, Erich Warsitz conducted high-speed taxi trials and, as was common practice, took the plane up a few feet over the runway.

Three days later, on August 27, the jet age was officially born as Warsitz lifted off from Marienehe and successfully flew ten minutes around the pattern before landing. This outstanding achievement was eclipsed five days later on September 1, at approximately 4:45 in the morning. Oberleutnant Bruno Dilley and a flight of Ju 87 Stukas came screaming out of the clouds and demolished the blockhouses on the Dirschau Bridge over Poland's Vistula River. Four German divisions simultaneously lunged across the border slamming into the unprepared and underequipped Polish defenders. Case White, the invasion of Poland had been executed, and with it the Second World War had begun.

Four

<hr/>

The Crucible

One hour past midnight on June 22, 1941, over 3,000 combat aircraft accompanied by seventeen panzer divisions punched huge gaps in the 1,800-mile Soviet border, and from the Black Sea to the Baltic nearly four million German soldiers cut through the shattered Red Army. On that June morning victory seemed a sure thing, more so by the end of the second day as 3,922 Russian aircraft had been destroyed against 78 losses for the Luftwaffe. This air superiority and the hard-charging panzers took the Germans 350 miles into the Soviet Union by the end of the first week. At the end of July, five Soviet armies and fifty complete divisions had been destroyed, with at least 600,000 men captured. If Hitler had stopped there, consolidated his gains, and negotiated a separate peace that included the oil- and mineral-rich Caucasus region, the war might well have been won. It was this ill-fated decision to continue east that eventually cost Hitler the war and destroyed his vaunted Thousand-Year Reich.

But he did not stop. The annihilation of Bolshevism, exter-
mination of Jews, and the enslavement of the Slavs were ideological dreams dear to Hitler and he believed, as he was wont to
do, that his military was unbeatable. Germany wanted the immense areas of living space and food available in the Ukraine
and desperately needed the vast resources and oil fields of Central Asia. Stabbing southward into this area would also threaten
Britain's Abadan oil fields in Iran, the world's largest source of
oil. Without it, Britain's war machine would grind to a halt and
peace in the west, under Germany's terms, very possible. This
was all within Hitler's reach and seemed quite feasible during
those heady months of 1941, which was one reason why various
jet programs were allowed to languish. Why was an advanced
aircraft necessary when Germany was winning everywhere
and now immeasurably aided, so Hitler believed, by Japan's
devastating December 7, 1941, attack on the U.S. Pacific Fleet
at Pearl Harbor?

Back in America the war seemed far away to George Welch
as he continued on to Purdue University to study mechanical
engineering. Joining the collegial fraternity life he thoroughly
enjoyed himself, but never lost sight of his goal to fly fighters.
Once he had the requisite two years of college behind him,
George applied for the Aviation Cadet program and was accepted into the Army Air Force just as Germany invaded Poland. A training slot opened up after the fall semester of his
junior year, and he happily left Purdue in December 1939 for
various ramshackle training bases and the uncertain life of
a student pilot. Surviving the four-phase training program at
bases in Texas and California, Welch emerged in January 1941
with about 200 hours of flying, the gold bars of a U.S. Army Air
Corps second lieutenant, and the prized silver wings of a pilot.

By February he was in Hawaii, on Oahu's Wheeler Field with the 47th Pursuit Squadron of the 15th Pursuit Group. There truly is no life like that of a bachelor lieutenant fighter pilot in a frontline squadron, and George made the most of the dances, the social scene, and the ladies. One story is told of a huge party at the Royal Hawaiian, a magnificent pink hotel on Waikiki Beach, to which he had no invitation. Deciding that his status as a fighter pilot merited an invite, he dressed in flight gear and swam offshore far enough to float past the beachside dance. Popping his parachute, George let the breeze tow him in toward the beach then calmly walked out of the surf into the party. The guests assumed he'd bailed out on a training flight, though he never admitted to that, and gave the young pilot all the drinks and female company he could manage.

During the first week in December, Welch and his squadron mate Ken Taylor were manning the dispersal strip at Haleiwa on the northern edge of Waialua Bay. Sharing tents with the bugs and sand was not how the pair intended to spend their weekend so on Friday, December 5, 1941, both pilots came down to Wheeler for a string of weekend parties. By Sunday they were hungover from days on the beach and nights at the officers' club. Poker, good drinks, and women had kept them up till the small hours of December 7, and nothing was going to get them out of bed.

Nothing except a Japanese sneak attack just after dawn.

By 0748 waves of Japanese Aichi D3A dive-bombers with Nakijima 97 torpedo planes were roaring over the island and sweeping into Pearl Harbor. The battleships were a primary target as Japan, and the United States to a large degree, were still unconvinced that aircraft carriers would rule the seas. The attackers concentrated on a sliver of shallow water between the

1010 Dock and Ford Island called Battleship Row. Airfields at Kaneohe Naval Air Station, Ford Island, and Wheeler were also high priorities to keep the surprised American fighter pilots from getting airborne. Blinking against the morning sun and their own headaches as they stumbled out onto the bachelor officers' quarter's huge patio, George Welch and Ken Taylor would have seen the "meatballs" painted on the attacking planes and sobered instantly. Not bothering with Wheeler, which was being bombed and strafed, the two pilots jumped into Taylor's new Buick convertible, peeled out, and headed to Haleiwa.

The P-40s there were full of fuel and .30-caliber ammo, and the ground crews already had the props turning. Half-dressed from the night before, Welch slipped into the cockpit and ran his eyes over the gauges. Beneath the big black and white turn indicator he read full fuel in the fuselage. Waving away the crew chief, George unlocked the tail wheel and with his right hand checked the cowl flaps. Men darted away beneath the wings pulling the chunky wheel chocks behind them and with his left hand Welch set the mixture to RICH, then slid the throttle smoothly forward.

By 0815 the two were airborne, gear and flaps up, guns armed, and Welch led the pair up over the little beach and away from Haleiwa, climbing slightly with wide eyes out. Training and instinct took over and George got his lap belt fastened, parachute straps connected, and adjusted the propeller pitch, all the while forcing himself to methodically scan the sky. Arcing around the west coast the pair of Army pilots caught their first sight of the enemy; Welch turned the reflector gunsight full up against the early morning glare, wincing as sunlight caught the mirror above the U-shaped crash pad. Shoving the mixture and throttle forward, the P-40s came roaring into the fight, ambushing twelve dive-bombers attacking the Marine Air Base at Ewa Field. Americans called the Aichi a "Val," and it was a slow, clumsy aircraft with fixed landing gear—dead meat against a fighter—and they each shot down a pair before landing at Wheeler to refuel and rearm. Astoundingly, the two pilots were berated by a staff officer for flying in nonregulation clothing; both men were wearing tuxedo pants under their khaki flying shirts. They were also told they had no authorization to fly and were ordered to stay on the ground.

Behaving as any real fighter pilot would do when faced with absurdity (and a battle raging overhead), Welch ignored the idiot and got airborne again as fast as possible. Though his Warhawk was shot up, George shot down another Val and a Mitsubishi "Zero" fighter just west of Barbers Point. Taylor also claimed another pair, which were eventually confirmed from Japanese records. For their gallantry that day both men were nominated for the Medal of Honor and received the Distinguished Service Cross, America's second-highest decoration for valor.* The entry for Welch's logbook that day only reads, "Combat patrol. The real McCoy."

Due to his actions at Pearl Harbor, George was sent back to the States for a war bond drive during the summer of 1942, and during this time the Japanese expanded their western perimeter from Manchuria into French Indochina (Vietnam) and Burma, south into the Malay Peninsula, Singapore, the Philippines, and all the way through the Solomon Islands to New Guinea. It was here, in early May 1942, that Japanese momen-

---

* The ridiculous downgrade was rumored to be as result of taking off without orders, and because they ignored the staff officer at Wheeler.

tum was finally slowed at the Battle of the Coral Sea. Japan was planning an invasion of Port Moresby, which would cut Allied supply lines and leave Australia open for invasion, but were met head-on by a small fleet built around two U.S. aircraft carriers: the *Lexington* and *Yorktown*.

Lieutenant Stanley "Swede" Vejtasa, a pilot aboard the *Yorktown* and future test pilot, shot down three Zeros with his dive-bomber on the last day of combat. Basically evenly matched, the actual battle was a draw favoring the Japanese, as the Americans were forced to scuttle the carrier *Lexington*. However, this was the first time in history that two fleets fought without the opposing surface ships gaining sight of each other, and naval warfare was changed forever. For those in any doubt, the Coral Sea irrevocably proved the strategic value and tactical necessity of carrier aviation. In the end, the proposed Japanese invasion was called off, Australia was safe, and after a decisive U.S. victory at the Battle of Midway in June, New Guinea remained relatively secure as a base for future operations.

August 1942 saw the 1st Marine Division come ashore at Guadalcanal and, after six months of bitter, vicious fighting they would annihilate the Japanese, turning the tide of the ground war as Midway had done it in the air. It was also in August that George Welch returned to combat. Finally escaping the publicity tours, speeches, and throngs of adoring women, he managed to get transferred back to a combat unit, the 36th Fighter Squadron of the 8th Fighter Group based at Milne Bay, New Guinea. Overjoyed to be back in the war, he was not at all happy about fighting the Japanese with the Bell Airacobra. Underpowered and awkward in many ways, the P-39D was so bad that the British Royal Air Force, despite their desperate need for fighters, had turned it down; nevertheless, the

Russians loved the plane and effectively used its big 37 mm cannon to kill German tanks.*

By late October, the Japanese navy was back in force in the southern Solomons to support a ground offensive that, it was hoped, would finally dislodge the tenacious U.S. Marines dug in on Guadalcanal. The Americans countered with another small fleet of two carriers under the command of Admiral William "Bull" Halsey. Swede Vejtasa, now in a fighter cockpit where he belonged, was part of Fighting Squadron (VF) 10 flying F4F Wildcats off the USS *Enterprise*. During a single mission on October 26, Swede shot down seven Japanese aircraft attacking his carrier and became the first American ace-in-a-day since the Great War.†

Throughout the fall of 1942, George flew and fought in defense of Port Moresby, and on December 7, 1942, shot down two Zeros and a Val dive-bomber on the anniversary of his first kill. According to his squadron mates, Welch often did not claim kills and was rather indifferent about keeping score. After all, he knew the truth, and being an inveterate prankster he enjoyed knowing what others did not. This included, if the stories are accurate, ditching a few P-39s in Milne Bay to get rid of them. The 36th Fighter Squadron pilots had been told to expect Lockheed P-38 Lightnings only when there were

---

* Four thousand seven hundred nineteen P-39s made it to Russia. About 1,100 were still flyable at the end of the war, and at least 1,030 had been shot down.

† Five torpedo bombers and two dive-bombers. America's first official ace-in-a-day was Lieutenant Frank Luke, U.S. Army Service, who shot down a Halberstadt reconnaissance plane, two Fokker D.VII fighters, and a pair of observation balloons on September 18, 1918.

no more Airacobras, so Welch and his buddies tried to speed the transition along.

This happened in May, finally, and Captain Welch transferred to the 80th Fighter Squadron on Port Moresby's Kila Airfield. During the summer of 1943 he shot down two more Zeros, and on a single August mission destroyed three Ki-61 fighters. The Hien, or "Tony" as the Allied called it, looked like a cross between an Allison-engined P-51 and Me 109. Sleek and powerfully armed, the Tony was a hard fight in the hands of an experienced pilot. In early September Welch gunned down a Ki-46 twin-engined reconnaissance plane and three more Zeros, bringing his total to sixteen confirmed kills—but he very likely had a half dozen more.

At that point life caught up with George. He actually airaborted a mission and returned to New Guinea due to a high fever, discovering that he had contracted malaria. Evacuated to the 118th General Hospital in Sydney, he recovered rapidly enough to troll for girls on nearby Palm Beach where he met a beautiful, suntanned Australian named Janet Williams. She, in turn, introduced him to an American paratrooper, a general named Joe Swing, commander of the 11th Airborne Division, and on his way into New Guinea. Despite "Jumpin Joe's" best efforts, Jan was utterly taken with the young fighter and on October 23, 1943, she married Major George Welch. After nearly two years of fighting and surviving 348 combat missions, he was preparing to return home just as another young American pilot was boarding the liner *Queen Elizabeth* in New York on his way to England: Charles Elwood "Chuck" Yeager and the 357th Fighter Group were going to war.

*   *   *

B ut the Axis momentum did not, could not, last.

By January 1943 the United States was well into the war and confident enough, even then, to host a conference in Casablanca, Morocco, where Germany's future unconditional surrender was already being planned. January also saw the capitulation of the German Sixth Army at Stalingrad, with a formal surrender occurring on February 2; it was an immense tactical defeat for Hitler and a stunning victory, both militarily and psychologically, for Stalin and the Allies.* The situation was dire enough that the Luftwaffe halted all training flights so precious fuel could be diverted to the Eastern Front. This had lasting implications as the flow of replacement pilots slowed, and those who did arrive lacked the superb training that characterized the prewar Luftwaffe.

Operation Torch, the Allied invasion of North Africa, had taken place in late 1942, and by New Year's Day the Japanese were about to lose Guadalcanal. The Axis had lost its momentum and, though in no way defeated, it was now known they could be beaten. Initially a shoestring offensive, and one that the United States was not really prepared to fight, Torch was nonetheless a strategic necessity. General (later Field Marshal) Erwin Rommel had been cleverly fighting in North Africa since late 1941, and by July 1942 his Afrika Korps had captured Mersa Metruh in Egypt. This put him within striking distance of Cairo, and if the city fell Rommel would certainly seize the Suez Canal, crippling Britain's lifeline to India.

---

* Ninety-one thousand out of 265,000 Germans survived to surrender, including 22 general officers. Less than 6,000 of these men would finally return to Germany in 1955.

An Allied offensive in North Africa would accomplish several immediate objectives. First, it would force Rommel to turn back from Egypt and fight the threat to his rear. This would keep the canal temporarily safe and forestall any moves by Egyptian nationalists who sought an alliance with Hitler. Allied control of the Mediterranean, or even a heavy presence there, would disrupt Rommel's main supply lines, cutting him off and effectively neutralizing the Afrika Korps. It would also get Lieutenant Ken Chilstrom, young fighter pilot and future demon chaser, into the war.

Allied landings in Morocco and Algeria would also sever the former French colonies from their collaborationist Vichy government, hopefully provoking a reaction by the French people and giving Hitler another significant strategic problem. Likewise, conquering Tunisia would be a severe blow to fascist Italy, sending a clear signal to those seeking Mussolini's downfall while providing a future base of operations against Sicily. An assault against the belly of the Third Reich would force Hitler to divert men and scarce resources to counter the threat. Such a riposte would also ease the pressure on the Eastern Front and give the hard-pressed Soviet Army some breathing room.

These were sound strategic concerns, so despite the U.S. military's reluctance, Torch was planned and put into action. Convoy UGF-1 departed Hampton Roads in the Chesapeake Bay on October 26 and was joined by a screen of warships, including the battleships *Texas* and *Massachusetts*, from Task Force 34 out of Maine's Casco Bay.* The aircraft carriers *Ranger, Chenango,*

---

* UGF: United States to Gibraltar, Fast convoy.

Suwanee, *Santee*, and *Sangamon* sailed from Bermuda and rendezvoused with the fleet as it headed for the African coast. The first of three prongs struck North Africa in Morocco before dawn on November 8, 1942, and the main objective for Major General George Patton's 33,000-man Western Force was the port of Casablanca.

It was believed, and planned for, that the French defenders would turn against the Germans and join the Allies, so there was no preliminary air or sea bombardment of the enemy coastal positions. This was a costly miscalculation as one surrender to the Germans in 1940 was enough for most of the French forces. Additionally, they had received intelligence of the Allied invasion fleets passing through Gibraltar headed for Algeria and were fully alerted. When the landings commenced, French shore batteries opened fire and their surface ships sortied, including several cruisers and five submarines. Vichy aircraft, among them Dewoitine D.520 and Curtiss Hawk Model 75 fighters, also got airborne in an attempt to repel the Allied invasion.

The 39,000-man Center Force sailed from the United Kingdom and came ashore in Algeria near Oran. In addition to the U.S. 1st Division, Major General Lloyd Fredendall had the 1st Armored Division, a battalion of paratroopers from the 509th Parachute Infantry Regiment, and the 1st Ranger Battalion. The mixed Anglo-American Eastern Force under U.S. Major General Charles Ryder hit the beaches on either side of Algiers. Resistance was light, and all the coastal batteries had been knocked out prior to the landings. Many of the French, including several commanders, immediately surrendered and welcomed the Allies.

In the end, Admiral François Darlan, commander of the Vichy navy, was captured in Algiers and, through a controversial deal with General Dwight Eisenhower, surrendered North Africa to the Allies. Though this outraged the Free French forces, who regarded Darlan as a willing Nazi collaborator (which he was), there were substantial strategic benefits. The Germans, who had never trusted the French, were compelled to initiate Case Anton, an operation they had foreseen in 1940, to occupy all of France. This meant that over twenty infantry divisions and a half-dozen armored divisions, desperately needed in Russia or North Africa, were not available. Less than two weeks after Darlan surrendered, the 7th Panzer Division rumbled into Toulon, the main port of the French fleet. Rather than be captured and refusing to steam to British ports, at least seventy-seven French ships were scuttled, including three battleships, seven cruisers, and twelve submarines.

Farther south the North African Atlantic ports of Casablanca, Safi, and Lyautey were now in American hands, as were all the surrounding airfields. U.S. losses were 174 killed, plus light damage to several ships, including the battleship *Massachusetts,* from shore batteries. Less than six aircraft had been lost, while Navy F4F Wildcats shot down over a dozen French fighters and transports. The Vichy battleship *Jean Bart* went down in Casablanca harbor and the French navy additionally lost a cruiser, four destroyers, and five submarines.

The 33rd Fighter Group flew their P-40 Warhawks off the escort carrier USS *Chenango,* landing at Port Lyautey on the newly christened Craw Field to begin combat operations.* Following

---

* While flying a white flag, Colonel Demas "Nick" Craw was killed by a French machine gunner as he attempted to negotiate the surrender of the Port Lyautey garrison.

the army advance, the 33rd would move up into Algeria on Christmas Eve 1942, first to Telergma Airfield then to Thelepte Air Base. By February 1943 Rommel was counterattacking U.S. forces at the Kasserine Pass and Thelepte was overrun. Shortly after this setback, the USS *Ranger* again arrived off Casablanca with three squadrons of P-40Ns from the 58th Fighter Group.[*]

"I was in the middle somewhere," Ken chuckles at the memory, "about number thirty-five or -six. And we didn't know until then that our brakes were no good." In February, they had flown the Warhawks from the Curtiss factory in Buffalo, New York, to the U.S. Naval Base at Norfolk. Once on the ground, the only way to get the fighters to the carrier was to taxi them over a mile to the docks. This wore out the brakes, but none of the pilots knew it was an issue until the chocks were pulled aboard the carrier weeks later. "We all drifted off the left side of the deck and sort of fell toward the water," Chilstrom recollected. "We all made it, though, and safely landed at Casablanca."

But it was there at Berrechid Airfield, which the fighter pilots promptly named "Bearshit Airfield," that they got the bad news.[†] Their P-40 fighters would go to the 33rd Group pilots because they had some combat experience. "They took our planes," Ken said ruefully. "So we sat around until a batch of A-36 Apaches arrived. They were shipped in by boat and covered with cosmoline grease. A depot group got them ready to fly

---

[*] The original P-40 was called the "Warhawk," and this name was officially used by all U.S. Army Air Corps units. Soviet and British squadrons called P-40B/C the "Tomahawk" until the D variant, or "Kittyhawk," became operational.

[†] About ten miles south of Casablanca on the A7, it is still in operation as Mohammed V International Airport.

and we took them up-country to Fez. It turned out to be a much better plane than the Warhawk."

And so it was.

The Apache was born from a private 1940 contract between the British government and North American Aviation (NAA) for 320 of the sleek, new fighters. The P-51/A-36 was the first mathematically designed aircraft in that every contour and every shape could be expressed algebraically. This resulted in extremely accurate, yet easily duplicated templates that were rapidly adapted for large-scale production, and this was essential for a big war effort, James "Dutch" Kindelberger correctly believed. A precisely constructed, mass-produced aircraft that was versatile in the air and easy to maintain in the field was a war winner. The Apache, which was the name NAA chose for the P-51, consisted of three main sections that could be disassembled in the field with an engine mount that required only four bolts and no special equipment.* The first prototype rolled out at Mines Field in Los Angeles on September 9, 1940, barely one hundred days after the contract was signed, and with it the pinnacle of piston-engined fighter development was in sight.

Though there were many significant aerodynamic points that matured as NAA's design evolved, it was the incorporation of a *laminar flow* wing that had lasting consequences as aircraft advanced into the jet age and men sought to conquer supersonic flight. "Laminar" essentially means "layered" and, as discussed, air flows over a wing in distinct layers. The one closest to the wing bonds to it molecularly to form a *boundary*

---

* The Royal Air Force, who purchased the initial planes, called it the "Mustang."

layer, like a thin coat of oil on a metal skin. If this boundary can be kept smooth and uninterrupted, then the layers above it will streamline and flow in smooth, regular paths. This substantially reduces drag, by as much as half in some cases, and correspondingly increases lift. Another highly significant result is greater speed, and once propulsion caught up to aerodynamics, anything, even flying faster than the speed of sound, was possible.

National Advisory Committee for Aeronautics (NACA) engineer Eastman Jacobs, head of the Variable Density Tunnel at Langley, designed such an airfoil that, based on only four digits, mathematically varied the thickness of a curve (camber) along the wing's chord line. Entire wings and airfoils of all types could now be studied and compiled into an aerodynamic database; "a classic—a designer's bible," J. D. Anderson writes about Jacobs's simple, yet pivotal method.* Laminar flow airfoils were symmetrical, but their maximum thickness was farther aft than that found on conventional wings, which meant at least 60 percent of the surface now produced a smooth, layered (laminar) flow, and increased lift. The science of it all made little difference to Ken Chilstrom and the others who flew the A-36 or P-51 in combat, but the practical results mattered enormously.

Laminar flow made the wing, and therefore the aircraft, very fast. Even in its infancy, the Mustang could sustain at least 375 miles per hour in level flight at 15,000 feet, making it about 30 miles per hour faster than the Spitfire. More critical still was the increase in range and flight endurance; the early A-36/P-51 could stay airborne four to five hours, and manage 1,000 miles

---

* *The Airplane: A History of Its Technology*, p. 231.

in range compared to the Spitfire's 400-mile range and two-hour endurance. This capability, and rapid improvements that followed with more powerful Rolls-Royce Merlin engines and better pilot training, now gave the Allies a true long-range fighter—a fighter that could roam at will over huge swaths of enemy territory and escort bombers on deep penetrations missions into the heart of the Third Reich.

These planes became part of the Allied Mediterranean Air Command, to which Ken Chilstrom belonged, which was formed under Air Chief Marshal Sir Arthur Tedder of the Royal Air Force. Its components included the Northwest African Tactical Air Forces (NATAF), the "teeth" of air operations in the Mediterranean, and this was commanded by USAAF Lieutenant General Carl "Tooey" Spaatz. Tactical operations involved purely air superiority fighter operations, which fell under the NATAF Desert Air Force with their P-40s and Spitfires. The deadly business of ground attack fell under the XII Air Support Command, composed of the 27th and 86th Fighter-Bomber Groups (so renamed in August 1943) with their A-36 Apaches, two groups of P-40 Warhawks, and the 31st Fighter Group—Americans like Bob Hoover flying the Spitfire Mk V. There was also the Strategic Air Force (NASAF), commanded by Major General Jimmy Doolittle, now a national hero after his April 1942 raid on Japan, and the Coastal Air Forces (NACAF), which were to shoot up any remaining Italian or German shipping they could find.*

---

* The Northwest African Photographic Reconnaissance Wing, also a component of NATAF, was commanded by Colonel Elliott Roosevelt, the president's son.

As for Ken Chilstrom, his immediate concerns were tactical, rather than strategic. After a month or so of A-36 Apache conversion training at Fez, Morocco, he was assigned to the 17th Light Bombardment Squadron of the newly reconstituted 27th Bombardment Group. Ken's unit forward deployed to Korba Airfield on Cape Bon, Tunisia, in May 1943, just as the remaining Axis forces in North Africa surrendered. There had been considerable debate within the Allied High Command and civilian leadership on the next course of action in the Mediterranean theater, and their subsequent strategy would put Ken into the fight.

The great island fortress of Sicily just off the Italian coast dominated the sea lanes and could threaten Allied North Africa so, if conquered, that threat could be removed and the Allies would gain a powerful base for operations into continental Europe. Hitler would have to answer the incursion, diverting men and resources from other theaters of war, and this would disrupt German progress with advanced technology like the jet. It was also felt that such an invasion might be the final straw to figuratively break Mussolini's back. There was serious Italian opposition to the war that had worsened all through the North African campaign, so with Tunisia lost and the Allies turning north, the Fascist dictator was in a truly precarious position. Mussolini was not trusted by his German allies, his own people were disillusioned, and his military, which was his power base, was faltering badly.

Winston Churchill, the British prime minister, wanted to bypass Sicily, cut it off, and invade mainland Italy. George C. Marshall, chief of staff for the U.S. Army and a superb strategist, wished to attack Sicily immediately. A fractured Italy could only hurt Hitler and aid the Allies, but he decided to begin by capturing the island of Pantelleria off the Tunisian coast with

Operation Corkscrew. Roughly halfway to Sicily, the forty-two-square-mile island was heavily fortified with over 100 emplaced guns of all calibers, including twenty-five heavy coastal pieces that would shred landing craft. Marshall had also learned that no aircraft carriers would be forthcoming to support his Sicily operation so Pantelleria's 5,000-foot-long Marghana Airfield, which could support eighty fighter aircraft, became an essential prize.

Doolittle's NASAF was wholly committed, including six groups of B-17 Flying Fortresses and B-25 Mitchell bombers. Corkscrew commenced on May 18, 1943, with over 100 bombing sorties each day pounding Pantelleria's guns, communications, roads, and harbor facilities. This continued for over two weeks and Ken Chilstrom flew into combat the first time on June 6, 1943, against the island's defenses, particularly those around Marghana. Slender and shark-like, his A-36 lifted off from Korba with the rest of the 17th Light Bombardment Squadron, and banked up sharply to the northeast. A vast, windswept gulf opened up off the nose, its foam flecked turquoise waters contrasting vividly against the mottled tans and browns of Cape Bon as land slid away beneath the wings. Left hand on the throttle and his right on the stick, Ken felt the Apache strain forward under the Allison engine's full power.

All around him the gray-blue sky was dotted with attack aircraft, but he focused solely on his flight lead. As the squadron rejoined, his eyes flickered between the other fighters and his cockpit. From the left side he saw his landing gear was up and above that the magnetic compass was swinging through north. From the corners of his eyes other Apaches appeared, all joining up in the briefed formation. Nudging the stick, he overbanked a bit to stay to the right of his leader then dropped his eyes straight down to the floor-mounted fuel selectors. They were both pointing right, positioned to feed gas from the fuselage. As the distance closed, the lead Apache was twice the size it had been a minute earlier, and the horizon seemed to rotate as the whole flight rolled out heading northeast and still climbing. He pulled the throttle back to hold position, and reactively scanned the two big gauges on the far right side of the panel. Tachometer was steady so the engine was fine . . . and the oil temperature was steady. Overheating could be problem during June in the Med with an Allison engine.

Not that there was time for that today. It was just over 50 miles from Korba to the squadron's target today . . . about thirteen minutes flying time at 225 miles per hour. The reflector gunsight was on full bright and with his left hand Ken flipped up the old-fashioned ring and bead sight next to it. Holding position 200 hundred feet off the leader's right wing, he then reached up under the glare shield and pulled the RIGHT HAND GUN charging handle, then the LEFT HAND GUN handle. Craning forward as the aircraft leveled off at 12,000 feet, he could see Pantelleria Island now . . . a dark green smudge against the water maybe twenty miles ahead. Ken had a map on his knee but didn't bother to glance at it.

He didn't need to.

The claw-shaped harbor on the northwest corner was plain to see . . . so was the mountain range that cut the little island in half. His target was Marghana airfield, smack in the middle of the flat plain between the mountain and the harbor. Even from here he could see the single east-west runway. Flashes . . . he caught flashes from around the airfield and harbor area, and just as he realized they were anti-aircraft guns, dark splotches suddenly appeared over the island.

So this is combat, he thought. Then the radios exploded with noise. Ignoring the chatter, Ken alternated between his leader and another scan of the instrument panel. Satisfied, he reached for a row of five toggle switches just left of the center pedestal. Flicking the first one up armed his guns and turned on the camera. Dropping his left hand back to the throttle, he pushed it up to stay in formation, then quickly flicked the third and fifth toggles up, which armed the nose fuses in his two 500-pound bombs.

Motion caught his eye and he saw the his leader surge ahead then waggle his wings. Dropping into trail formation, Ken rolled up slightly to keep the airfield in sight. Airfield . . . leader . . . airfield. There were the guns . . . those were the squadron's targets. The wind was blowing from the west so they would hit the revetments on the east side first so the dust and smoke wouldn't obscure the other targets. Ken was to hit anything on the far east edge of the field and south of the runway. His leader would hit anything north and the rest of the squadron would work back to the west.

Crossing the shoreline, he could plainly see white specks all over the dark green landscape . . . homes and buildings. There! A service road bulged south of the runway and right in the middle was a cluster of revetments. Black spots sprouted everywhere around him, but Ken ignored them and tensed as the big latticed dive brakes on his leader's wings opened up and the Apache rolled onto its back. It seemed to hang there for a long moment, then vapor streamed from the wingtips as the pilot pulled straight down.

Thousand one . . . thousand two . . . he forced himself to

count slowly then rolled inverted. Hanging upside down, he stared at the twin peaks two miles below him then popped his dive brakes, and felt the Apache shudder. Instantly jamming the throttle forward, he blinked, spotted the runway and revetments, then, as airbursts exploded all around, pulled straight down into the frantically firing guns.

"They [the Italians] weren't very good," he recalls. "They were more lovers than fighters." Ken and the rest of the squadron clobbered the target then headed southwest back to Korba, low and fast over the Gulf of Hammamet. By the time the British 1st Infantry Division came ashore on the morning of June 11, 5,285 sorties had been flown and 6,200 tons of bombs dropped. Over half of the island's AAA guns had been knocked out along with most of the coastal artillery, the power plant, and all eighty Italian aircraft. The demoralized and shocked Italian garrison surrendered without a fight, and the only Allied casualty was a British soldier who had been bitten by an agitated donkey. By July 10, Marghana was operational and temporarily home to the 33rd Fighter Group as the Allies commenced Operation Husky, and the invasion of Sicily. Unlike Operation Torch, there were no concerns in Sicily about the defender, and extensive preinvasion "softening up" took place during the preceding two months.

Over 42,000 combat sorties were flown, and the 27th Group was in the thick of it, flying surface attack missions against the Italian and German defensive positions. There were ports, roads, marshalling yards, and nineteen major air bases to be hit; most critical among them was the Gerbini complex of fields on the southeastern side of the island. Trapani, Palermo, and, most vital, the port of Messina, were all consistently and heavily attacked. The A-36 proved a killer in the lethal world of close air support; its long loiter time, six .50-caliber Brownings, and two 500-pound bombs put the Apache in high demand as Sicily was relentlessly hammered prior to the invasion. Ken Chilstrom and his personal plane, which he named *Little Stinker*, flew in four-ship formations of two pairs apiece. Dive-bombing from 14,000 feet, they would easily exceed 400 mph during the attacks, and pulling off target with their dive brakes closed, they reached 450 mph thanks to the laminar flow wings.

Operation Husky consisted of two main assaults and, as with Torch before, it was excellent practice for the coming invasion of France. Elements of the 82nd Airborne were dropped behind Gela between the Hermann Göring Division and the coast. The idea was for them to slow any counterattack and hold on until a linkup occurred with the seaborne infantry. Paratrooper Joe Swing, before heading to the South Pacific, had planned the daring move. Task Force 545, composed of the British Eighth Army and 1st Canadian Infantry Division, landed on the southeastern edge of the island between Syracuse and Pachino. They were to move north on the Catania coastal road past Mount Etna toward Messina and deal with the Italian Napoli Division along the way. About fifty miles to the west the American Seventh Army under George Patton came ashore along the Gulf of Gela. The idea was to cut the island in two pieces, with the Americans blocking any Axis reinforcements from Palermo in the west while protecting the British left flank. Encountering the Italian Livorno Division and the Hermann Göring Division in the mountains beyond Gela, the Americans managed to break out by July 12.

On July 18, the 27th Bombardment Group moved to Gela Airfield and Ken continued flying close air support missions.

sometimes several per day, interdicting Axis reinforcements and generally shooting anything that moved. That same day Hitler was in Italy to encourage, lecture, and scold a despondent Mussolini and stiffen Italy's resistance. Due to heavy losses on the Eastern Front, a generally poor showing in their North African backyard, and the loss of "our sea," as the Fascists called the Mediterranean, Italy had been put dangerously close to a regime change.

The Germans never trusted their Italian allies, but they did regard them as a buffer against any assault into the southern Reich and, true to form, had several plans waiting for implementation should Italy turn. For their part, Italians felt slighted by the contemptuous treatment they often received at the hands of their Axis ally, and they resented the perceived lack of German support in defending Sicily. This point was intentionally driven home on July 19, as some 500 Allied bombers dropped 1,100 tons of bombs on Rome. Over the next seventy-eight days 110,000 sorties would be flown against the "Eternal City," though the Vatican was carefully spared.

Back on Sicily, Ken Chilstrom was heavily involved in the U.S. breakout to the west. Patton, being Patton, did not accept a secondary support role to the British Eighth Army, so he reorganized his army into separate forces and attacked in three directions. While Task Force 545 was bogged down south of Mount Etna, the U.S. 2nd Armored completed a stunning end run around the far western tip of the island. Chilstrom and three other A-36 pilots, all lieutenants, were in a jeep one day near Gela when a staff car suddenly blocked the road in front of them. Two officers got out, including a general wearing riding pants and cavalry boots. It was Patton himself, and he was not happy. "Where are your fucking helmets?" he growled.

"Don't you assholes know any better?" Seventy-four years later Ken still smiles over the incident, though it wasn't funny at the time. "We used helmets for washing and shaving," he said, shaking his head. "What good are they in a fighter plane?"

Palermo fell to the Allies on July 22, 1943, and it was here that Ken got a short break when Bob Hope and the Gypsies arrived with the USO. "He was a real gentleman," Chilstrom fondly remembers. "He hung around after the show, talking and ad-libbing. I sat next to him at dinner and for an hour could forget the war." Hope respected all the fighting men, but especially pilots. "One of the aviators here took me for a plane ride this afternoon," Hope was fond of recounting. "I wasn't frightened, but at two thousand feet one of my goose pimples bailed out." *TIME* magazine agreed and wrote of the legendary entertainer:

> *Like most legends, it represents measurable qualities in a kind of mystical blend. Hope was funny, treating hordes of soldiers to roars of laughter. He was friendly—ate with servicemen, drank with them, read their doggerel, listened to their songs. He was indefatigable, running himself ragged with five, six, seven shows a day. He was figurative—the straight link with home, the radio voice that for years had filled the living room and that in foreign parts called up its image. Hence boys whom Hope might entertain for an hour awaited him for weeks.*

But it could not last, so Ken was back at war the next day and Sicily was far from a certain victory. Most of the island's defenders, including the 15th Panzergrenadiers and the Hermann Göring Division, fell back and dug in with their backs to

Messina, the only port left open to them. This defensive pocket, called the Etna Line, was also the area where most Axis fighter opposition was encountered as it was very close to Luftwaffe bases in southern Italy.

Then on July 25, 1943, a major fracture appeared in the Axis wall. Mussolini, officially the prime minister of Italy, received a "no confidence" vote by his own Grand Fascist Council and was ordered to resign by King Victor Emmanuel III. Defeated and despondent, Il Duce was promptly arrested and would spend the next six weeks being shuttled around Italy to prevent his rescue by the Germans. "We hoped this would make the fight for Sicily a bit easier," Ken remembered, "but it seemed to have no effect on what we were doing." This was certainly true regarding the German military. In fact, the situation became more dangerous as many Italian units melted away into the local area, which left the Germans isolated and desperately vicious.

In the midst of this, Operation Tidal Wave, a massive B-24 strike launched from Benghazi in North Africa, aimed to cripple output from the Romanian refineries clustered around Ploiești. Though not operating at full capacity, these facilities provided at least 30 percent of the Reich's oil and were extremely valuable targets. One hundred sixty-two B-24 Liberators from five bombardment groups made it into Romania on August 1, 1943.* Hit hard, production was indeed crippled— for a few weeks. The damage was quickly repaired and output actually increased, yet the costly Allied raid unequivocally ex-

* The Eighth Air Force's 98th and 376th Bombardment Groups and the Fifteenth Air Force's 44th, 93rd, and 389th Bombardment Groups.

posed Germany's Achilles' heel and proved without question
that even the best-defended targets in the Reich were vulnerable.* It also graphically revealed the absolute necessity of a long-
range fighter to protect the bombers.

All through July Ken Chilstrom and his squadron hammered
at Sicily's defenders. Montgomery's British Eighth Army was
still battling the Hermann Göring Division along the coast road
near Catania, but by the end of the month Patton was in Santo
Stefano approaching the Etna Line from the west through the
Sicilian Apennines. The Germans knew they could not hold on
and Field Marshal Albert Kesselring had already decided on a
fighting withdrawal, in carefully orchestrated stages, across the
Strait of Messina to Calabria on the Italian mainland.

Chilstrom and the rest of the fighter-bombers could do nothing about it. At its narrowest point, the strait was less than two
miles across to the Via San Giovanni, and with hundreds of
anti-aircraft artillery (AAA) guns on both sides it was one of
the most heavily defended areas of Europe, certainly in the
Mediterranean, because the Germans knew it was their only
way out of Sicily. "We lost some folks there," Ken remembered.
"One friend of mine, Harry Castleman, went down somewhere
around there. A British Walrus [Air-Sea Rescue] found his raft
but no Harry. We thought he was just gone, like so many guys
who go down over water. Turns out"—he smiled—"Harry
couldn't get to his raft so he swam ashore. He ended up in a
convent and hid out from the Germans until we landed in Italy.

---

* Fifty-three bombers were shot down and only eight-eight made it back to
their home bases.

Two months after he went down he strolled into the squadron! He looked great. Guess the nuns took good care of him."

The Calabrian side of the strait bristled with coastal artillery, including two 170 mm batteries, and the latent threat of the Italian navy was always present. Impressively, the Germans managed to evacuate about 8,000 men per night using carefully controlled routes and ferry crossings. In the end, through a clever shifting of troops, minefields, and obstacles, 100,000 Axis troops, including the two panzergrenadier divisions and the Hermann Göring Division, got clean away. They took with them 14,000 vehicles, 47 tanks, and nearly 100 heavy guns that would now face the Allies in Italy.

Sicily had fallen, and with it Mussolini, at the cost of nearly 23,000 American, British, and Canadian soldiers killed, wounded, or missing. German losses stood at 27,940, while over 150,000 Italians were captured, were killed, or went missing. With the capture of Sicily the strategic situation changed drastically. The Mediterranean became untenable for the Axis, and sea lanes from Gibraltar to Cairo were now open to transport Allied men and matériel anywhere across North Africa, up to Greece, and east into the Levant. Vital supplies could also reach the Soviet Union through the Dardanelles and Black Sea.

The loss definitely hastened Mussolini's fall and Field Marshal Pietro Badoglio became prime minister of a country caught on the proverbial anvil: a German advance from the north, and an imminent Allied invasion from the south. Ken Chilstrom got a few days of rest, but the situation was extremely fluid, especially after the Italians concluded a separate peace in early September with the Armistice of Cassibile. That same day, elements of the British XIII Corps under Montgomery crossed the Strait of Messina and landed in Calabria—Axis Europe had been invaded.

Bob Hoover, now part of the 4th Fighter Squadron, was flying Spitfires out of Palermo, shooting up transports and German warships operating from the French coast. Two days later Chilstrom and the 27th Group moved up to the Barcelona Landing Ground in Milazzo Harbor, a few miles southwest of Messina. The Italians attempted to gain favorable conditions under the armistice but were in no position to bargain. When the treaty was announced on September 8, 1943, the Germans predictably reacted harshly and moved to rapidly occupy the rest of Italy. The Italian air force effectively ceased to exist and was of no consequence to the Allies, but the *Regia Marina*, the Royal Navy, was another matter. As a protocol of the armistice the Allies had insisted the fleet not be scuttled or handed over to the Germans, who were moving to quickly seize the main ports of Genoa, Taranto, and La Spezia.*

What happened next arguably ushered in the age of precision-guided munitions. Admiral Carlo Bergamini slipped out of Genoa aboard the battleship *Roma* at 2:30 A.M. on September 9, after telling the Germans he was going to attack the Allied landing sites near Salerno. In company with the battleships *Vittorio Veneto, Italia*, and a dozen escorts, he initially headed for Sardinia before finding the port of La Maddelena had just been seized by the Germans. Heading for Malta, the fleet was caught transiting the Strait of Bonifacio between Corsica and

---

* Most of the fleet sortied and made it to Malta or North Africa.

Sardinia and attacked by eleven twin-engined Do 217 bombers out of Marseilles.*

The Germans used the Fritz X, a 3,450-pound bomb guided by radio control link, to attack the battleship. This was a manual control to line-of-sight (MCLOS) system so the Dornier's bombardier relied on flares mounted on the bomb's tail assembly to keep sight. He would then use a Kehl transmitter to transmit guidance signals to a Strasbourg receiver, and through this the pitch-and-roll spoilers on the tail were activated to guide the bomb to impact. Designed as an antiship weapon, the Fritz was made to be dropped between 18,000 and 20,000 feet. Its 705-pound warhead could penetrate up to twenty-eight inches of steel plate, though warship deck armor was much thinner. The bombers began attacking about 1530 as the fleet approached Asinara Island but were too high for the anti-aircraft guns to reach. *Roma* was struck and survivors later noted that the bombs appeared to follow the ship as it maneuvered. By 1552, drifting and afire, the battleship was hit again, but this time the massive bomb penetrated the forward engine room and from there into the magazine. The resulting explosion killed Admiral Bergamini as well as the ship's captain; twenty minutes later the 46,000-ton warship broke in two, capsized, and sank with 1,253 of her 1,849 man crew.

The same day the *Roma* went down, the Allies invaded mainland Italy at Salerno, twenty-five miles south of Naples. Ken Chilstrom and the 27th Fighter-Bomber Group were once

---

* Gruppe III of Kampfgeschwader 100. This was the only operational unit to use the Fritz X.

again in the thick of things.* Operation Avalanche was intended to cut off Axis forces south of Naples and, by seizing that port, ensure a solid, deepwater base for resupply and future operations. The British Eighth Army was to drive north from Calabria, mopping up resistance along the way, then link up with the U.S. Fifth Corps at Salerno. The combined force would then advance north toward Rome.

It was a bad plan.

First, based on flawed assumptions from the German evacuation of Sicily, the Allies assumed the Germans would not fight for Italy but retreat north. It was also not recognized that the Germans regarded Italy as a buffer against the southern Reich, and they intended to bleed the Allies every step of the way. Second, supplies and reinforcements for the Italian campaign were a lower priority due to the buildup in England for the impending Allied invasion of France. Last, in a silly attempt to achieve surprise, there was again no pre-assault bombardment ordered for Avalanche. The Germans obviously expected an invasion and had multiple divisions in the landing area, including those that had escaped from Sicily.

The 16th Panzer Division was commanded by Rudolph Sieckenius, a veteran of France and Russia, who formed four mobile battle groups to oppose the landing. Allied air cover, including Ken Chilstrom, had a tough time against the panzers because they deployed in small units of five to seven tanks, and

---

* On August 23, 1943, the 27th Bombardment Group was redesignated the 27th Fighter-Bomber Group, and the 17th Bombardment Squadron was renamed the 523 Fighter-Bomber Squadron.

only concentrated air support and heavy naval gunfire from the cruisers *Philadelphia* and *Savannah* halted the counterattack. "It was touch and go from the beginning, very confusing . . . and no one was certain of the situation from hour to hour." Chilstrom was adamant about the resolve, however. "We weren't going to let the Germans push our guys back into the sea," yet they very nearly did. Taking advantage of the unfavorable terrain on the Salerno plains, the Germans attacked through the open, unoccupied gap between the Sele and Calore rivers, and came within a mile of the beaches on the morning of September 9. "As sure as God lives the Germans will attack down that river," Patton wrote from Sicily, and he was quite correct.

Counterattacks continued for the next two days as the Germans threw at least six divisions, under strength but veteran, into the battle. On September 13, two panzer divisions and the 29th Panzergrenadiers nearly made it to the beaches, but they temporarily halted on the Calore River because the bridge was destroyed. It was here that American field artillery zeroed in, an intense concentration of 155 mm and 105 mm howitzers that literally stopped the panzers in their tracks. By the end of the next day, under continuous air cover, most of the U.S. 45th Infantry Division was ashore at Paestum, on the southern end of the Gulf of Salerno, and the U.S. 3rd Infantry Division was crossing from Sicily.

The Germans were now relearning a lesson they'd been taught at El Alamein by the British, at Stalingrad by the Red Army, and now at Salerno by the Americans—if their initial counterattack was not successful, then there likely would not be a chance for another. The open ground exploited by the panzers now became a kill zone as 500 American medium and heavy bombers blasted the Sele-Calore corridor. The USS *Philadelphia* and *Boise* got close inshore near the mouth of the river and opened fire at point-blank range on anything that moved. Farther offshore, the Royal Navy battleships HMS *Valiant* and *Warspite* pounded the inland areas with their sixteen-inch guns. Overhead, Apaches and other fighter-bombers hit bridges, roads, and anything German they could find.

Regrouping one final time, the Hermann Göring Division and 26th Panzer attacked the Salerno beachhead on September 16 and were stopped cold. Two days later Ken Chilstrom and the 27th Fighter-Bomber Group moved 150 miles north from Milazzo to Capaccio Airfield just outside Paestum. From here the 523rd and other squadrons ranged along the coast, bombing and strafing as the Germans pulled back to the north. Operation Avalanche was successful in that the Allies were ashore in strength and could not be dislodged, yet it failed because Naples was not captured; the Allies now faced a long, bitter advance up the peninsula, and the Germans were in no way vanquished from Italy.

In fact, during the fighting around Salerno, Mussolini was rescued from the Gran Sasso area of Abruzzo in central Italy. Waffen SS commandos and regular paratroopers under the command of Major Otto Harald-Mors snatched the former dictator from the Campo Imperatore ski resort and he was eventually installed as head of northern Italy's Salò Republic, though he was never more than a figurehead and the Germans remained firmly in control.* Yet a protracted war of attrition as

---

* Oberleutnant Georg Freiherr von Berlepsch of Fallschirmjäger-Lehr Bataillon actually freed Mussolini. This is often and incorrectly attributed

the Allies prepared to invade France was something the Third Reich could ill afford, and that is exactly what they had. Both sides also had a recalcitrant, unpredictable, and generally hostile Italian population to control.

Chilstrom would move again in early November 1943 to Guado Air Base near Bellizzi, and it was here that he would finish out his tour in Italy. Nothing, no school or training program, can hone flying skills like combat; through the crucible of 1943 Ken had acquired a veteran's judgment and experience that would serve him well for the rest of his career. He developed controlled aggressiveness; that vital, hard-won and unteachable attribute that distinctly marks a successful combat fighter pilot. He'd killed tanks and scores of vehicles; dropped bridges; and shot up the German and Italian armies—but air combat was scarce. The Italians were out of the picture, not that they had challenged Allied fighters much, and most of the Germans were still on the Eastern Front or defending France.

Finally, on his seventy-third mission, Ken was in a position to add an enemy plane or two to his list. While leading a flight of four Apaches into a valley west of Rome he spotted three Junkers floatplanes on a lake.* "I called 'power back' to the rest of my flight. We've got to let 'em get up so we can shoot them down." Weaving back and forth just over the ridgeline, the four fighters waited, but the Germans never took off. "I think they

---

to Otto Skorzeny, who was present and did escort the former dictator out of Italy, but did not plan or lead the raid.

* Ju 52 trimotor, nicknamed the "Iron Annie," it could fly from Italy to Berlin in about eight hours.

saw us and decided they'd live longer if they stayed on the water. We finally ran out of gas and had to shoot them up where they floated. I got the first one." He still smiles at that memory.

Other memories were not so good.

On another mission Chilstrom led a flight of eight Apaches into a valley east of Mount Maiella in the rugged Chieti region to hit some gun emplacements. He did everything right: a medium-altitude reconnaissance, and an attack with the sun behind him in the eyes of anyone looking up from the valley. Suddenly he caught glimpse of a blue flare. "The Germans had learned." He shook his head slowly. "They would send up a flare like that as a warning whenever A-36s were around . . . but we were already in the valley. Before we could get the hell out, five Apaches went down in less than a minute. The coolant lines and radiator of the A-36 were particularly vulnerable to ground fire and once they were damaged the fighter would not last long." The other three tried to get away from that valley but had to bail out . . . they were all captured. "The Germans, at least these Germans, were very considerate about alerting us to who became prisoners of war."

Allied victories in North Africa, an invasion of Italy, and a disastrous campaign in Russia were clear warnings to those in the Reich who could still read the writing on the wall. Even if Hitler refused to accept the strategic situation, there were others who did. As the Allied bombing campaign was beginning to take hold and the Wehrmacht in Italy was inexorably pushed northward, German technical advances offered a solution for an increasingly desperate military situation. Short of a negotiated peace, which was highly unlikely, Germany's wonder weapons and superior technology seemed the only way out.

Five

======

Wonder

Adolf Busemann and Alexander Lippisch were worried men.

As with many academics, they had far-reaching collaborative interests that transcended national borders, and contacts in a wide variety of countries. Until World War II, that is. For several years the conflict had forced them to work in a vacuum without the benefit of scientific cross-fertilization that often occurs in academia, but, as long as Germany was winning, the situation was tolerable. Yet that had all changed by 1943 and both men, leading aerodynamicists, had cause for concern—especially with the Russians closing in from the east. The demand for expertise was so dire that the German government recalled scientists, engineers, and technicians from combat duty. Ordered back into research and development, which they should never have left, over 4,000 rocket specialists alone returned to Peenemünde, Oberammergau, and other top-secret facilities.

Lippisch, a Bavarian by birth, had served as an aerial observer with the Imperial German Air Service from 1915 to 1918, and after the Great War worked for the Zeppelin Company. Eventually completing a doctorate in engineering from the University of Heidelberg, Lippisch was an advocate of supersonic flight while much of the world believed it to be fantastical nonsense. Through experimentation and wind tunnel tests, he became convinced that with sufficient power a delta wing design (basically variations on a triangle) would permit flight faster than sound. His colleague, fellow engineer Adolf Busemann, was similarly convinced that such speeds were attainable, but he advocated a different method. Busemann, born in Lübeck along Germany's northern coast, had been part of a gilded development team that included such legends as Theodore von Kármán and Ludwig Prandtl, and his specialty was airflow; specifically supersonic airflow.

Not particularly a new field, supersonic flow had been researched for years to improve the efficiency of steam turbine engines. "I worked as an engineer and had learned in college, of course, about steam turbines and things like that," Busemann reflected during a 1979 interview. "They were already invented. Therefore, we wanted to see how to make them most efficient—how to get the most energy out, put the least into a reversed flow, and reduce energy losses.

It was through his expertise in this area that Busemann developed the concept of a *swept* wing: one set at an angle less than 90 degrees to the fuselage. This had been done before as early as 1908, and J. W. Dunne, a British engineer, built a flying-wing biplane that made it across the English Channel in 1913. But Dunne and several others who used swept wings were concerned with pilot visibility and longitudinal stability, while Busemann was thinking in terms of speed.

The problem was air.

Or rather, how air reacts as a body moves through it at velocities approaching the local speed of sound. Remember that this varies with altitude as the higher one ascends, the lower the temperature; for standard conditions at sea level a body is supersonic at 1,117 feet per second, while 49,000 up to the speed of sound is less: about 968 feet per second. Amateurs and engineers alike, specifically ballistics engineers interested in improving artillery shells, understood that as a body approached this speed the drag acting against it increased. This observation was deliberately noted by Benjamin Robins, an accomplished British military engineer, in an article published in 1742.

But it was Ernst Waldfried Mach who first gave a numerical value for velocity as it related to the speed of sound; this was expressed in tenths, up to the whole number 1.0, which represented (and still does) supersonic flight. Mach was an Austrian, born in 1838 to a cultured, intellectual family who tutored him from an early age in geometry, algebra, and science. Mentally agile, Mach's natural abilities in philosophy, music, medicine, and languages stimulated his creativity into many diverse fields.* Earning a PhD in physics from the University of Vienna, Mach served as a professor of experimental physics at the University of Prague for nearly three decades.

Then, in 1887, he saw the demon itself and, astonishingly, photographed it. A precise perfectionist and superlative experimentalist, Mach set up trip wires triggered by a passing

---

* American author Marilyn vos Savant, listed in *Guinness World Records* as the holder of the "Highest IQ" between 1986 and 1989, is a descendant of Ernst Mach.

bullet and then photographed the event. Well aware that air is disturbed by a body's movement through it, he also knew that such a disturbance would refract light. Even in a transparent medium such as air, such a refraction would cause shadows, like sunlight striking waves, and these could be captured by photography—which they were. (The image appears in this book's photo insert.)

Revealed to the Academy of Sciences in Vienna, the image of the shock waves surrounding the bullet was staggering. The bow wave is clearly visible just ahead of the bullet as was the turbulent, churned-up air in the wake. This point, Mach 1 as it came to be known in 1929, was where a body—projectile or plane—became supersonic. A further refinement became known as the *Critical Mach Number*, the lowest speed that some point on an aircraft becomes supersonic, but does not exceed it.

Mathematical calculations for supersonic flight were relatively straightforward, and now there was visible evidence of the shock wave, but very little was understood about air as it approached Mach 1. Termed the *transonic* region, this was a murky, ill-defined area where flows are locally lower or higher than the aircraft's forward velocity. So while the trailing edge of a wing might be subsonic, the leading edge, for example, may have exceeded the speed of sound. Air behaved differently in these regions, and there were considerable problems, especially in a world of straight-winged aircraft.

Below the speed of sound, air is essentially incompressible (for our purposes) and cannot be "packed" any tighter than it is. In this region, a body or an aircraft disturbs the surrounding air, splits it, and sends out pressure pulses like a bow wave from a boat moving through water. As long as the aircraft remains subsonic there are no aerodynamic issues, and the flow remains relatively smooth with predictable actions. Now, we know that as air accelerates, its pressure changes, and as long as the speed remains well below the speed of sound, it does little more than generate the lift needed to fly. But when a body, or part of a body, accelerates into the transonic region, then different areas of it are now subject to erratic and variable pressures that subject the body to unplanned effects. The most severe of these was a slew of new aerodynamic challenges collectively termed *compressibility*.

Some of these had been identified as early as the Great War by engineers seeking to improve propellers, and it was well known at the time that even at a modest 130 mph aircraft velocity, the propeller tip speeds were well past the speed of sound. Essentially, air becomes "thinner" as it speeds up and the molecules spread out, which does a lot of bad things, aerodynamically. Pressure decreases, which simultaneously increases drag and decreases lift. This was confirmed during a series of experiments conducted under the auspices of the NACA by Hugh Dryden and Lyman Briggs of the National Bureau of Standards in the late 1920s. Courtesy of the General Electric Company, they improvised a wind tunnel and ran a series of physical tests to validate earlier American and British research.

Briggs and Lyman definitively proved that thicker wings at higher angles of attack were susceptible to transonic effects at lower speeds than a thin wing, or one at a lower angle. In addition to lift and drag, they also concluded that the center of pressure on a wing moved aft toward the trailing edge as speed increased. This was significant, though not realized at the time, because it would directly and often fatally affect controllability since the ailerons were located on the trailing edge. Also, remembering Mach's photograph, that turbulent wake would subsequently flow over an aircraft's horizontal tail and impact elevator effectiveness. Yet what could be done about it?

Air cannot be altered, so the answer had to lie with altering the effects air had upon a wing. In fact, the theme of the 1935 Volta Conference in Rome was "High Velocities in Aviation," and specifically the properties of subsonic to supersonic airflows. Hosted by the Royal Academy of Science, the leading aerodynamicists in the world convened in Rome during September 1935. Among them were Gaetano Arturo Crocco, Eastman Jacobs of the NACA, Hugh Dryden of the Bureau of Standards, Theodore von Kármán, and Jakob Ackeret, who first articulated the "Mach Number" in 1929 in deference to Ernst Mach. Compressibility and its effects took center stage and Eastman Jacobs with John Stack of the NACA were the leading authorities on the subject. In addition to the precise, unambiguous findings, Eastman Jacobs also had a series of schlieren photographs, very much like Mach's shadowgraph, which visually depicted the effects of the "compressibility burble," a term coined by Stack.*

Adolf Busemann was certain he had the solution. A wing's aerodynamic characteristics are dictated by a component of the airflow's velocity perpendicular to its leading edge, so if the angle of the wing is decreased to less than 90 degrees (a straight wing), then the component striking the leading edge will also decrease. This would change everything. The critical Mach number would be higher, so the wing would fly faster before

---

* This type of photography illuminates an object in a flow field, using light to capture shadows formed by the differences in density that result from pressure changes as a body approaches and exceeds supersonic speeds.

becoming supersonic, and the huge drag coefficient associated with this would be delayed. Even when it did occur, the severity of the increase would be much reduced. With a powerful engine, like a jet, the door could be opened for supersonic flight.

It is interesting to speculate how the air war over Germany might have played out if the potential of the swept wing had been grasped by the Allies in the mid-1930s. It is also tantalizing to picture the mating of Frank Whittle's and General Electric's jet engine research to a swept-wing aircraft in 1942, especially with the vast money and resources available to the Allies.

The presentation of the swept wing and a solution to this nasty aerodynamic dilemma should have been a godsend but, amazingly, it was virtually ignored. Those in attendance were focused primarily on theory, not design, and the evolutionary breakthrough was treated rather indifferently by von Kármán and others. The Germans, on the other hand, took it very seriously. Busemann confirmed his swept-wing data through high-speed wind tunnel testing in 1939 at Braunschweig, and this was subsequently used by Messerschmitt for development of the Me 262 and the next technical leap forward: Projekt 1101.

It was apparent to these men that the transonic region was the real danger, and if an aircraft could not be controlled as it transited the Mach, then it would not physically survive to get through to the other side. The speed of sound was just a number, an artificial flag created by men and not a barrier at all—compressibility was the true gateway into the demon's world.

Paradoxically, even as battlefield reverses accelerated German advanced technology development it had the opposite effect on the Allied jet program. With tactical and strategic successes, and in the absence of real threat, the Allies, particularly the Americans, rightly concentrated on perfecting and fielding proven technology that worked, rather than dissipating their efforts as the Germans did. Nevertheless, Major General Hap Arnold's 1941 visit to Britain convinced him (and he convinced the U.S. government) that the technology gap must be closed. Just over ninety days prior to the attack on Pearl Harbor, the General Electric Company of Schenectady, New York, was contracted to build the initial American jet engine: the GE I-A.

In a surprising move, Bell Aircraft was asked to design the aircraft around General Electric's engine, and the contract was signed on the last day of September 1941. The company's efforts with the P-39 Airacobra and P-63 Kingcobra had been average, at best, and Bell's reputation for building effective, frontline fighters was questionable. Yet they had a tolerance for unconventional ideas, and perhaps this was the reason behind the choice or, more likely, Bell's Buffalo, New York, facility was the closest to General Electric, and this would help ensure secrecy for the project. In any case, the XP-59, America's first jet program, was alive.*

By late summer 1942, as the Marines were landing on Guadalcanal, the first prototype was ready—but where to test it? Wright Field, though close to the engine and airframe manufacturers, was deemed too populated not only for secrecy but for safety in case anything went wrong. What was needed was a

---

* It was labeled XP-59 as a cover. There was an existing piston-engined XP-50 program so if discovered, it was hoped the Axis would simply believe it to be another conventional fighter.

place no one would look, no big cities nearby and very little civilization. A place that was easy to secure, far from prying eyes,
and with good enough weather to permit nearly constant flying. A place where nothing on the ground would be destroyed
by falling aircraft because there was nothing on the ground to
destroy. Such a place did indeed exist: Rogers Dry Lake in Antelope Valley, California.

Situated on the roof of the Mojave Desert, the lake bed was
close enough to Los Angeles for convenience, but separated
from the coast by the San Gabriel mountain range. It offered
120-degree temperatures, almost no rain, scorpions, flies, and
snakes; combined with Rosamond Dry Lake, it also offered approximately 306,000 acres of flat, dry landing surfaces. Originally a water stop for the Santa Fe Railroad, the Rodriguez
Mining Company had controlled much of the land, but in 1910,
a few years after the Wright Brothers flew, Clifford and Ralph
Corum arrived. The land was free, and the government was so
desperate to attract settlers that it offered a $1 per acre incentive
for every acre that was improved. The Corums called their new
home "Rod," short for Rodriguez, which means "Roger" in English, and the dry lake now had an Anglicized name.

The Corums went into business attracting other homestead-
ers to the area and helping them drill wells, clear the land, and
ship in supplies. A general store was built, and also a church.
To encourage growth and legitimize the little community, Ef-
fie Corum, Clifford's wife, petitioned the U.S. government to
name the post office after her family, but this was denied as a
Coram, California, already existed and the similarity was too
close. The Corums then reversed their name and suggested
"Muroc," which was accepted, and the area now had an official
name.

In 1933 the Army Air Force, which had a penchant for deso-
late areas and decrepit, threadbare bases, sent a small detachment out from March Field to design a bombing and gunnery
range. For the next eight years this was manned by a handful
of Army personnel and used for aircrew tactical training while
the government quietly bought up all the surrounding land it
could. After the Japanese attack on Pearl Harbor, units of the
41st Bombardment Group left Davis-Monthan in Tucson for
Army Air Base, Muroc Lake, as it was called. Joined later by
the 30th Bombardment Group, the population of the dusty, obscure post went from dozens to thousands within days, and in
1943 was renamed Muroc Army Air Field.

Primarily utilized as an operational training base, Muroc
and its satellite airfields existed to put the final touches on all
sorts of pilots heading into combat: bomber crews in B-24 and
B-25s; P-38 Lightning fighter pilots, and A-20 attack pilots. The
ramshackle collection of buildings on the southern shore of
the dry lake, called "South Base," continued to grow. Bombing
and strafing were the focus, and to that end the Army built a
650-foot-long mock-up of a Japanese *Takao*-class heavy cruiser.*

Muroc was perfect.

Yet by 1942 it was recognized that a remote test site was
needed for the Army's more exotic programs. Wright Field in
Ohio and Florida's Pinecastle were both becoming too populated, so a small highly classified annex was constructed on
the north side of Rogers Dry Lake. Officially known as the Ma-

---

* Officially designated as "T-799 Japanese Battleship, Plan No 944/41
W-509-Eng 4239" and was used until 1950. Then, in true Army fashion, this
ship in the desert was labeled a "hazard to flight" and dismantled—which
took some time due to the pile of unexploded bombs in the belly.

terial Center Flight Test Site, the program engineers, technicians, and contractors called it "North Base," and secrecy was so tight that the Army personnel on the south side initially had no clue what was happening. The XP-59 was taken west by rail in September, its jet engine covered with a tarp and a dummy propeller on the nose to allay suspicion. Robert Stanley, Bell Aircraft's chief test pilot, first got the first XP-59A Airacomet (#42-108784) into the air from Muroc on October 1, 1942. It was underpowered and, like all jets of the day, prone to overheating. Cockpit visibility was poor, as was its acceleration compared to frontline piston-engined fighters.

Service test versions, now the YP-59A, were delivered to the military in June 1943, and even with 1,650 pounds of thrust each from more powerful GE I-16 engines the jet was still a dog. Despite its lackluster performer, controllability issues, and marginal engine response, the U.S. government ordered eighty of the new jets, though this number was eventually cut in half as the military correctly resisted expending time, money, and resources on unproven and, as they saw it at the time, unnecessary technology. Certainly the American attitude would have been different had the Luftwaffe fielded a jet fighter in sufficient numbers and early enough in the war to make a difference, but it had not. Yet there were those in Washington who could see, even in 1943 and 1944, that the next threat to emerge from the war would not be from Germany or Japan. The next threat would be as much a clash of ideals as of technology and it would be through technology, not numbers alone, that peace would be maintained.

Nevertheless, German technology was still fearful, and guided bombs were not the only new technology to appear during the summer of 1943. Wild rumors began circulating about strange German fighters with no propellers, and speeds no Allied plane could match. Some pilots were aware of jet engine technology, and some were not. It was certainly not new, nor was it particularly secret unless connected to a specific aircraft program. "We weren't really worried about it," Chilstrom recalls. "The Germans were at least six years ahead of us in this regard, and yet they hadn't been able to put many into operation. From what I knew about jets they wouldn't be much good at low-altitude operations . . . the engines weren't made for it. And low-altitude ground attack was my life in 1943."

The same month Ken Chilstrom entered combat during June 1943, Frank Whittle and Gloster's E.28 second flying test jet logged over fifty hours in the air. Rolls-Royce had taken over production of the engine and Britain expected to field an operational jet fighter in 1944. The United States Army Air Force, thanks to a Whittle engine given to General Electric in 1942 and with vastly more resources available, rapidly caught up to the Royal Air Force.

June 1943 also saw Lockheed acceptance of a new government contract, and the legendary Kelly Johnson delivered his own jet fighter design proposal. Work commenced on the XP-80 just before the Allied invasion of Sicily and, realizing they were beginning at a disadvantage, Lockheed opted to build its own fuselage around an existing British jet engine: the Halford H-1B. Much of the initial design came from plans for a single-engine version of the XP-59 that Bell had provided. Johnson and his talented design team, the famous "Skunk Works," delivered their aircraft body in November 1943 after just 143 days.

Larger and heavier, the straight-winged jet sported tricycle landing gear and mounted the engine *inside* the fuselage, rather than in external nacelles. The new General Electric I-40 engine produced 4,000 pounds of thrust, more than double that of the I-16 in Bell's Airacomet. To test and evaluate both jets, the Army Air Force formed the 412th Fighter Group, America's first jet unit, and based it on the north edge of Muroc's dry lake. Though the "Shooting Star," as it would be called, did not fly until early 1944, it would outperform Bell's Airacomet in every way, and become America's first operational jet fighter, though too late for the Second World War.*

Though waning German fortunes of war accelerated the development and fielding of the Luftwaffe's advanced aircraft programs, these efforts were too widely dispersed for real efficiency, especially under the current conditions. Still, they had not been idle, and were still well ahead in jet and rocket technology. Piloted by Heini Dittmar, the Messerschmitt Me 163 Komet flew in 1941, and the initial preproduction models were delivered to Erprobungskommando 16 (Service Test Unit 16, to EK 16), in July 1943, under the command of Major Wolfgang Späte, a ninety-nine-victory ace. Capable of flight past 600 miles per hour, the stubby little interceptor would remain the fastest manned aircraft of the war, and its technology would influence the future design of the Bell X-1.

Using detachable, dolly-type landing gear the Komet would roar off the ground and rocket up toward the belly of heavy bomber squadrons at an astonishing 11,800 feet per minute.

---

* It had issues of its own, though, especially the fuel pump system, and this resulted in crashes that killed Milo Burcham, Lockheed's chief engineering test pilot, and Major Dick Bong, America's leading ace.

Once in range of its two Mk-108 30 mm cannons, the pilot would open fire, then shoot vertically up through the bombers much faster than gunners or escorts could react. Apexing about 10,000 feet above the formation, the Me 163 would then dive down through them, firing again at the pass. Engagements were limited as the little interceptor had a short range, and eight minutes of fuel at best so once this was burned up the pilot would glide back to land. But the Komet was a rocket; true, it could fly extremely fast and hit hard, but it could never dogfight, perform close air support, or be taken seriously in an air superiority role at a time of flight measured in minutes.

What was needed was a real fighter, one that could do several missions, be mass-produced, and sustainable in combat. The Germans, specifically Messerschmitt, catapulted into the future by pairing a successful engine to a swept-wing fighter design with Projekt 1065; the result was the Me 262 and the world's first operational jet fighter. Scientists such as Hans von Ohain, Adolf Busemann, and Woldemar Voigt had remained enthusiastic about the possibility before the war, and after the conflict began, the National Air Ministry (RLM) could see its obvious potential for air combat.

Heinkel had the obvious lead as the He 178 jet had flown in 1939. Its successor, the He 280 prototype, was completed during the 1940 Battle for France and first flew in March 1941 about the time George Welch was floating ashore on the Royal Hawaiian beach at Waikiki. The later-model Heinkel featured tricycle landing gear and the world's first ejection seat on a production aircraft. But despite initial success, including outdogfighting a Fw 190, Heinkel never enjoyed Messerschmitt's favor with the Luftwaffe and RLM.

After three years of research and development, Fritz Wendel took the initial jet-powered Me 262 off the ground from Leipheim on July 18, 1942, three months prior to Bell's Airacomet and nine months prior to the Gloster Meteor. In April 1943, while Ken Chilstrom was learning the A-36 in Morocco, Hauptmann Wolfgang Späte first flew the fighter for the Luftwaffe. Cross-purposes at the top of the Nazi hierarchy, Allied bombings, the German penchant for endless tinkering, and the pronounced lack of engines seriously delayed fielding of the Me 262. Adolf Galland was one of those pushing hard for a thousand jets per month. He wrote:

"The problem which the Americans have set the fighter arm is . . . quite simply the problem of superiority in the air. The enemy's proficiency in action is extraordinarily high and the technical accomplishment of his aircraft so outstanding that all we can say is something must be done! We are numerically inferior, and always will be . . . I am convinced that we can do wonders even with a small number of greatly superior aircraft like the Me 262 or Me 163."

Even then its effectiveness was largely negated by Hitler himself. Convinced (rightly so) that the Allies would invade northern Europe, Hitler insisted that the Me 262 be employed as an unstoppable "Blitz" bomber best employed for lightning-fast strikes against the invasion fleet. The first operational deliveries of the 262 came in June 1944, as by the time the converted Me 262A-2a, known as the Sturmvogel (Stormbird), became operational with Kampfgeschwader 51, the Allies had landed and were themselves blitzing across France. Yet even in less than a year of operational flying, the 262 claimed over 500 Allied aircraft, and there is long-standing debate on what could have happened if the jet had been fielded just a year earlier.

Had the early technical advantages been seriously exploited,

developed, and pursued from 1939 on, the air war over Occupied Europe might have been very different indeed, but as we
have seen, by 1943 the Germans had already lost the war. That
fall there were 109 operational Me jets; 70 Blitz bombers, and
39 pure fighter versions. If the air-to-ground variant managed
a monthly 2:1 kill ratio and the fighter maintained 3:1, then this
would inflict a hypothetical Allied loss of 335 aircraft per month
to the jet. During combat in one month alone, April 1944, the
USAAF in Europe lost 683 aircraft and this was *before* the Me
262 became fully operational. According to the *Army Air Forces
Statistical Digest*, a total of 27,694 aircraft were lost during
World War II; 8,481 fighters, 8,314 heavy bombers, and 1,623
light or medium bombers—and this was just the Army.

Fielding the jet a year earlier would not have had a significant tactical impact, and due to Allied bombing, shortage of
scarce materials, and transportation issues, the Me 262 could
not have been mass-produced during 1944 in sufficient numbers to overcome the sheer numbers of Allied aircraft. Also,
at least 25 percent of the 564 jets produced in 1944 were not
accepted as fit for combat. Late that year Hitler again changed
his mind and decreed all jets would now be fighters. Due to the
delays, alterations, and politics, only 1,433 Me 262s would ever
be produced and barely 100 would see action at any given time.
The Allies were certainly aware of the jet and, after an early
mission where six jets shot down fifteen heavy bombers in a
matter of minutes, were justifiably wary.

Yet the Americans had purposely chosen another stratagem
until the war was won. Their leaders correctly decided that
U.S. research would focus on revising and perfecting existing
aircraft designs, which were all created prior to Washington's
declaration of war on December 7, 1941. So the time, funds,

and resources were to be expended on mass-producing *existing* airplanes rather than chasing phantoms. It worked, obviously, because the Allies did achieve air superiority and won the war, but had Germany not invaded Russia in 1941 and devoted the resources to fielding a jet in 1942 before America could effectively intervene, the war, and history, could have well gone another way.

Six

The Brave

*S*trawberry Bitch, Aluminum Overcast, F-Bomb, Cocktail *Hour, Boobs Not Bombs, Reluctant Dragon*: and thousands more. American B-17s of the Eighth Air Force's 97th Bomb Group arrived at High Wycombe, England, during May 1942, and the first heavy bomber combat mission was launched on August 17 against a French rail yard near Rouen-Sotteville. Over the next three years Allied bombers would drop some 1.7 million tons of bombs, 70 percent of the total dropped on the Reich, on strategic facilities like ports, rail yards, factories in Germany and France. At least 10,000 missions would be flown during the next three years and 47,000 flyers, roughly the equivalent of three U.S. Army divisions, would be lost in combat, yet these casualties did not make the Allies consider a negotiated settlement, nor did German defenders gain enough time to effectively field Hitler's last resort: the weapons of the Wunderwaffe. These included poison gas delivery systems for tabun and sarin, drones, electromagnetic cannons, and directed energy weapons that induced magnetic-interference-stalled aircraft motors. From AEG Siemensstadt in Berlin to GEMA-Werke in Silesia (and dozens more like them) potentially lethal wonder weapons were in various stages of completion all over the Reich.

Yet it was the psychological effects that held a particular fascination for Hitler, who could never forget his devastating World War I experiences from artillery. In fact, the term "psychological warfare" likely originated from *Weltanschauungskrieg*, or "worldview warfare." An early example were the Jericho sirens mounted on the B-1 variant of the Ju-87 "Stuka," but as long as Germany was winning the war such ideas were interesting notions at best.

Two events changed all this: the Battle of Britain and the invasion of the Soviet Union.

By September 1940 it was obvious that the Luftwaffe had failed to gain air superiority over the channel, and Sea Lion, the invasion of England, could never go forward without it. Battling the Royal Air Force was decidedly different from fighting the valiant but outclassed Poles, or the dazed, ineffective French Armée de l'Air. In the five months between the fall of Dunkirk and the suspension of Sea Lion, the Germans lost at least 2,870 aircraft of all types and over 4,000 superbly trained, experienced aviators. Glaring technical, tactical, and strategic weaknesses within the Luftwaffe had been revealed. There were no long-range fighters or heavy bombers—they weren't needed for what had been developed primarily as a tactical, close-air-support type of air force. Yet to pound British ports, factories, or radar installations into submission this is exactly what was required, and it was the lack of such aircraft that provided the impetus for the initial vengeance weapon: the V-1 rocket, and its successor, the V-2.

The rocket did not require air superiority, did not need fighter protection, and could, if used en masse, inflict considerable damage against critical targets. Conceived by Dr. Fritz Gosslau of Argus Motoren, the 27-foot, 4,750-pound V-1 was first launched under its own power on Christmas Eve 1942. The Allies were well aware of the guided weapons project, called Vulcan, and specifically targeted its known production facilities. On a moonlit night in mid-August 1943, nearly 600 RAF heavy bombers struck the Peenemünde Army Research Center in northern Germany.* Though this and other raids delayed V-1 operations, the first rocket hit London in the East End on June 13, 1944, impacting near Grove Road between Victoria Park and Canary Wharf.

Numerous defensive measures were taken, which included barrage balloons, decoys, and the employment of fighters as interceptors though the piston-engined fighters were rarely quick enough to catch the missiles. Cheap and lightweight, the V-1 used a simple pulse jet engine that, though primitive, unquestionably demonstrated Germany's technical lead. Once ignited, a pulse jet squirts fuel into a simple combustion chamber and ignites it. The resulting blast, or pulse, of energy compresses through a narrow aperture. This results in highly accelerated exhaust that creates thrust. A pulse jet did not require magnetos or igniters and would burn any type of petroleum, even low grades captured from the Soviets.

---

* 324 Lancasters; 218 Halifaxes; 54 Stirlings. Forty bombers were lost.

Launched originally from fixed sites along the Pas-de-Calais, the rocket reached speeds in excess of 400 miles per hour and could fly its 1,870-pound warhead to London in about twenty-five minutes. By mid-June, 500 flying bombs had been dispatched, though the Normandy landings necessitated a shift of launch sites to the north. But the Luftwaffe had another nasty surprise for the Allies in the form of the world's first combat jet fighter: the sleek, swept-winged Schwalbe. First blood was probably drawn on July 26 when an Me 262 damaged a photo-reconnaissance RAF Mosquito Mk. XVI from 540 Squadron.[*] The next day several Gloster Meteors from 616 Squadron, Britain's first operational jet unit, were in action over Kent against the flying bombs. Possessing the speed to catch the V-1, the Meteor and Hawker Tempests destroyed some 80 percent of the incoming missiles.

With the lurking potential of the Wunderwaffen, the operational employment of jet aircraft, and missiles came the modern age of aerial warfare. It had definitely begun, albeit on a small scale—and it had to be stopped. The Allied solution, really the only effective response possible, was total air superiority. Control of the skies meant the risk of invading Europe could realistically be taken and a foothold then gained, and held, on the Nazi-occupied continent. Air superiority meant regular, dependable supplies, and unfettered movement of matériel and troops to conduct offensive action on the Allies' terms. This would drive the Reich into a truly defensive, reactive posture and eventually force an unconditional surrender that would

---

[*] Claimed by Leutnant Alfred Schreiber of Erprobungskommando 262 flying an Me 262 A-1a.

end the war. With air superiority all this was possible—and without it none of it would happen.

The Royal Air Force did not risk its few Meteors in air-to-air combat, and its Spitfires and Tempests, though exceedingly fine aircraft, had relatively short combat ranges.* The American jet program had yet to field an operational aircraft so the burden of air superiority over the Reich fell to long-range USAAF conventional aircraft: the P-47 Thunderbolt, the P-38 Lightning, and, above all, the North American P-51 Mustang. They were named *Slender, Tender, and Tall*; *Iron Ass*; *Glamorous Glen*; *Grim Reaper*; *She Wouldn't Wait*; *Passion Wagon*; and hundreds of other, deeply personal, funny, or inspirational names chosen by its pilots. To the rest of the world they were Mustangs; more specifically, the Merlin-engined North American P-51D.

Unquestionably the pinnacle of piston-engined fighter design, the Mustang was the quintessential propeller-driven fighter. Every contour was a derivative from a geometrical shape; beautiful, to be sure, but this also meant the blueprints could be expressed algebraically and all associated templates easily mass-produced. Initially a private venture between North American Aviation and the British Air Ministry for the NA-73X prototype, an initial 320 aircraft contract was signed on May 29, 1940. A scant 102 days later the first Allison-engined Mustang rolled out at Mines Field, Los Angeles, on September 9, 1940—two days after the bombing of London began.†

---

* Tempest V at approximately 740 miles and the Spitfire Mk. XIVe at 460 miles.

† The Luftwaffe had inadvertently bombed London in late August but September 7 marked the first of fifty-seven consecutive nights of bombing

The culmination of hard-learned combat lessons, the Mustang's flexible design allowed continuous evolution against changing threats. Visibility from the beautiful bubble canopy was superb and, unlike the Spitfire or Bf 109, the P-51's landing gear retracted inboard toward the fuselage, maintaining a heavier center of gravity and permitting thinner wings. With a wide, twelve-foot wheel base, operations on marginally prepared forward strips were much safer. Pilot inputs were incorporated into the cockpit layout, and the result was a relatively roomy design with all the essential switches, gauges, and controls in sensible locations—thus freeing the pilot up to fly and fight. Yet for all its deadly beauty the Mustang, along with every other piston-engined fighter, could only fly so fast or turn so hard. This was due to the inability of even the most powerful piston engines to overcome the aerodynamic drag forces acting on the aircraft as it approached transonic speeds.

During dives they did get closer to the speed of sound than aircraft ever had before, but at their own peril. Lockheed test pilot Ralph Virden died in late 1941 when he could not pull his P-38 out of a high-speed dive, and the culprit was compressibility. Test pilots discovered during testing, and combat pilots in combat, that the P-51 could achieve about 0.84 Mach before it was in danger of going out of control. Due to its thinner wings, a Spitfire had actually reached 0.92 Mach in a dive. Measuring the local speed of sound, that is, based on the current altitude, temperature, and pressure, was not a new idea. Austrian physicist Ernst Mach had conceived of the value during the nineteenth century, and modern aerodynamic engineers knew that

known as the "Blitz."

exceeding the speed of sound in an aircraft was theoretically possible, especially with rocket motors or jet engines.

Supersonic flight and jets were exciting and, most believed, they were the future. However, neither would end the conflict, and winning World War II remained the focus of the Allied governments. Oil storage, refining, and distribution targets were the priority, for without oil the entire Nazi war machine would die of thirst. Wartime requirements were seven to eight million barrels per month, and this was cut to the bone without any large-scale, offensive action. The Allies knew the German weakness even before the war and estimated that if enemy oil production could be reduced by 50 percent, then the Reich would fracture. This Achilles' heel was one strategic reason behind Operation Barbarossa—the invasion of Russia— and Hitler's second fatal error.

Besides oil, by late 1944 bombing had disrupted the advanced aircraft programs and production of the wonder weapons. It had completely shut down the manufacturing of submarines, particularly the deadly Type XXI and Type XIII U-boats. Ammunition production was drastically impacted to the point where field units rarely had enough bullets or mortar rounds and the virtual nonexistence of nitrates resulted in many artillery shells being packed with rock salt. Transport vehicle production was similarly devastated; Daimler-Benz, a huge manufacturer of vehicles, engines, and spare parts, was effectively wiped out during raids on Stuttgart. Opel, Germany's largest maker of trucks, was bombed to bits during a series of raids on its Brandenburg factory. Ford Motors had developed interests in Germany since the mid-1920s, and through its factory in Cologne it made Maultier half-tracks for the German army and turbines for V-2 rocket motors. Incredibly, after the war both General Motors, which had owned Opel since 1931, and the Ford Motor Company both sued the U.S. government for wartime damages to its German-based businesses.*

Critics of the bombing offensive often point to its failures, if they can be called that. Steel production was minimally impacted and conventional aviation manufacturing actually increased due to superb German damage control and reconstruction efforts. Some 50 to 60 percent of the Reich's ball bearings, essential for machines of all types, were manufactured by the Svenska Kullagerfabriken (SKF) plant at Schweinfurt, and the main facility in Göteborg, Sweden. Schweinfurt was attacked multiple times between August and October 1943 and, during a single mission on August 17, 1943, sixty B-17s were shot down with at least eighty more irreparably damaged.

Without ball bearings the flow of critical war matériel throughout the Reich would grind to a halt.† Schweinfurt was heavily defended by over 100 anti-aircraft guns, including the lethal Flak 18 and 36 types, at least nine big 150 cm searchlights and artificial fog generators. Eighth Air Force raids slowed production by a drastic 30 to 35 percent, yet the shortfall was immediately made good by enormous reserves within the Reich, SKF Sweden, and extra shipments from its largest plant—in Philadelphia.

Reaching a wartime peak of about 1,800 per month in De-

---

* General Motors was awarded over $30 million while Ford received $1.1 million.

† Four thousand bearings were used in each Focke-Wulf 190 fighter.

cember 1944, tanks and armored vehicles continued to roll off assembly lines, and aircraft production was also generally unaffected by the bombing. In 1941 German factories had produced 360 fighters per month, but under Erhard Milch, the Reich's air minister, this increased to 1,000 per month by mid-1943. Fifty percent of these fighters were made in only two locations: Regensburg in Bavaria, which made the Bf 109, and the Wiener Neustädter Flugzeugwerk outside of Vienna.

Yet even with the might of the Eighth Air Force pounding away at it, materially the Luftwaffe was not truly hampered by Allied bombs until early 1944. Albert Speer admitted in his memoir that until 1944 nothing had been destroyed that could not be rebuilt, but the constant combat cost the Luftwaffe its irreplaceable fighter pilots. Aircraft could be manufactured, and eventually peaked at about 1,500 fighters per month, but pilots were a different matter altogether. It took eighteen to twenty years to raise a man, and the Reich didn't have that much time. The opening months of 1944 cost the Luftwaffe 1,684 fighter pilots, including nearly 60 *experten*: aces. So critical was the shortfall that young men began arriving at frontline units with less than half the flying time of their Allied adversaries, and they rarely survived their first few missions.

However, Allied strategic planners rightly concluded that without oil and logistical transport capabilities, then the tanks and planes would eventually pile up in depots—and they did. By the spring of 1944 there were other competing priorities: invasion targets and the wonder weapons. Infrastructure, namely bridges, rolling stock, and canals, was pulverized. This made even limited resupply efforts tenuous at best and prevented rapid German counterattacks following the Allied invasion of June 1944. Railways, namely bridges and marshalling yards, were primary objectives and over 100 such targets were identi-fied in the Low Countries and France.

The Germans had made good use of rolling stock from the defeated continental nations since they shared a common nar-row gauge (4' 8½") rail, and like the Deutsche Reichsban, all the French, Dutch, and Belgian locomotives ran on coal, which was one resource Germany possessed in abundance.* So much so, that early jet fuel was made from a lignite derivative. Loco-motives were especially lucrative targets because without them nothing would move, so hundreds of them were destroyed by roaming Allied fighters. By D-day on June 6, 1944, it was reck-oned that the entire French transportation system, essential for German combat operations, was operating at less than 60 per-cent of capacity, and the risk of effective German counterattack was greatly lessened. Albert Speer, Hitler's armament's minis-ter, later wrote that the bombing campaign "meant the end of German armaments production" and he estimated 98 percent of the Reich's oil production capacity had been lost by July 1944.

Bombing did not have that catastrophic effect on other facets of the Reich economy, but—and this is a vital and oft-overlooked point—the bombings, by necessity, did force a reaction from the Germans. Paralyzed by aerial assaults and now beset on all sides, the remaining realists in the Reich were aware that the war could not be conventionally concluded with satisfactory terms. In light of all this, July saw the Jägernotprogramm, the Emergency Fighter Program, revealed. Aircraft production, and what resources remained, were to be focused on defensive inter-ceptors to stop, or at least slow, the Allied bombing campaign.

---

* Spain and the Soviet Union utilized wide, or broad gauge, railways.

Luftwaffe generals, Adolf Galland among them, opposed the program as a waste of what few resources Germany possessed, and as not viable in the presence of such overwhelming Allied air superiority. Overruled by Hermann Göring and Hitler, over forty different ramjet, turbojet, and rocket interceptor designs were created between July 1944 and the spring of 1945, one of these being Bachem's Natter. Simple, inexpensive construction was normal for most of these programs and, as with the Japanese strategy, these were aircraft that could be flown by minimally trained, fanatical youngsters. Human life, it seemed, was cheap in defense of the Reich. Interestingly, all five of the final designs from Messerschmitt, Heinkel, Blohm &Voss, Junkers, and Focke-Wulf incorporated *swept wings*, the practical value of which had yet to be fully realized by the British or Americans.

Allied bombings during the winter of 1944 forced nearly thirty primary aircraft plants to disperse into 700 smaller factories, with engine and parts manufacturers doing the same. For a nation dreadfully short of resources this complicated matters considerably. The overtaxed labor pool struggled to keep up, quality suffered from lack of concentrated talent, and the inadequate transportation network buckled under constant Allied attacks. Some aircraft assembly plants relocated to forest factories (*waldwerken*) and straight sections of autobahn were used as runways.* The forest option was attractive because it was cheap; a *waldwerke* in Gauting cost less, could contain up to 1,200 workers, and was virtually impossible to locate from the air. In fact, none were bombed nor was their existence even

---

* The Kuno 1 *waldwerke* and the A8 autobahn near Augsburg was used for the Me 262 Messerschmitt jet fighter.

known to Allied intelligence. By comparison the tunnel factory option cost five times as much with half the capacity.

Fortunately for the Allies, German fascination with the unique persisted till the bitter end. The Esche II complex near Sankt Georgen in northern Austria was enormous; six miles of concrete-reinforced tunnels that managed to produce 987 Me 262 fuselages by the end of the war. Another method, popular in the occupied territories, was to pour an immense concrete slab, let it harden, then dig out the earth beneath and reinforce as necessary. Quick, virtually free, and effective, this "earth mould" construction was intended for V-series weapon sites within range of Allied bombers.

But it was too little and far too late.

Believing that a dispersion of critical factories would reduce output, which it would but nothing like being bombed to pieces, Germany committed another fatal strategic blunder. Nazi propagandists were also certain that to suggest the Reich could be badly damaged from the air was a defeatist attitude and, in 1942, this certainly appeared true. The RAF bombings till that point had targeted cities and population centers, not production facilities or infrastructure. This was in part due to the British belief that burning out cities was demoralizing and would strike at Hitler's base of support—the German people. However, the attacks were also carried out in piecemeal fashion because the RAF lacked the sheer numbers of bombers and fighter escorts to make daylight bombing even remotely feasible. This would have to wait until the fall of 1942, and the arrival of the B-17 and B-24 heavy bombers of the American Eighth Air Force.

To counter the onslaught, the Luftwaffe fought back hard. From 1940 to 1944, across multiple fronts, fighter aircraft of all kinds remained operational with an *average* monthly attrition of 10 percent or less. Pilots fully trained in peacetime who had gained combat experience in Poland, Scandinavia, and France made up the bulk of the fighter arm, but losses against Britain's Royal Air Force and the 1941 invasion of the Soviet Union took a heavy toll. By 1944 what remained were a very few hardened and nearly unbeatable *experten*, and many poorly trained replacements. Yet time was running out for them as well, and December 1944 alone would see 452 pilots killed or captured. Adolf Galland, the charismatic and deadly *General der Gagdflieger* (General of Fighters) for the Luftwaffe, would later write that the bombing was "the most important of the combined factors which brought about the collapse of Germany."

And it was collapsing.

The average German lived on bread or potatoes and received a single bar of soap each month; an egg per week was wild good fortune. Sugar, coffee, and any type of meat went to the military. Fresh vegetables, fruit, or fish were nonexistent and many government salaries were paid with coupons for coal or milk. What food remained was largely produced by women, children, and nearly one million French prisoners of war. Horses all but disappeared as they were used by the Wehrmacht as draft animals or for food. The situation was desperate on the ground. During the first six months of 1944, while the Americans finalized plans for their X-1 supersonic flight research project, 1.5 million German soldiers went missing, were captured, or were killed trying to slow the Red Army's advance from the east.

By late summer and early fall of 1944 the Allies had landed at Anzio and were moving through Italy. Fortress Europe had been penetrated by Operation Overlord and the Allied landings in Normandy. The Wehrmacht had been pushed back as far as the Dutch border and, on August 25, 1944, the German garrison in Paris surrendered. Unsurprisingly, despite the tanks, rifles, uniforms, fuel, and transport back into France that had been provided by the Americans—and in direct defiance of orders from the U.S. commander—a French unit slipped into Paris first to "liberate" the city. Somewhat ironically, it was French in name only.* Nevertheless, by mid-September the Germans were falling back everywhere; in the center Patton's Third Army had fought its way into the Alsace-Lorraine, and in the south the U.S. Seventh Army pushed north from Toulon into the Rhône valley.

The battle for air superiority over the Reich was dire for the Luftwaffe, and by now over three-fourths of all its fighters were engaged directly against the U.S. Army Air Corps or Royal Air Force. Long-range American fighters, particularly the P-51 Mustang, could now escort bombers all the way to Berlin, and after D-day the forward-based RAF Tempests and Spitfires were roaming into Germany itself. In fact, it was over Berlin on March 4, 1944, that Pilot Officer Yeager bagged his first Bf 109 after three weeks of combat flying with the "Yoxford Boys"; the 363rd Fighter Squadron. His luck changed the next day when a Focke-Wulf 190 shot his Mustang to pieces, forcing Chuck to bail out over southern France. Managing to escape and evade, Yeager was taken over the Pyrenees into Spain by the French Resistance and was back in England by the middle of May.

---

* Nicknamed La Nueve, this was the Ninth Company, Regiment of the March of Chad, and was composed of Spanish soldiers commanded by a French captain.

Bob Hoover, who once said "I'd rather dogfight than eat steak!," had indeed gotten his wish. Based out of Calvi, Corsica, he had downed an Fw 190 before suffering the misfortune of meeting Luftwaffe veteran Siegfried Lemke on February 9, 1944. Hoover was one of four Spitfires the German claimed that day, and he went down twenty miles off the French coast. Numb from the cold and freezing, Bob was captured by a German corvette and eventually ended up in Stalag Luft 1. Gus Lundquist had also finally made it to the war in the summer of 1944, flying P-51s out of England. Ironically, given his test work with the Fw 190, he was shot down over France in July by a Focke-Wulf. Also an inmate of Stalag Luft 1, Lundquist taught Bob Hoover everything he knew about the Fw 190, which would come in very handy later in the war.

Though backed into a corner in the summer of 1944 the Luftwaffe was still a fierce opponent, now made desperate in defense of its own country. Consequently, the German High Command pulled nearly all aviation assets off the Eastern Front to defend the Reich from the western assault; some 1,560 fighters, over 80 percent of the total available aircraft, vainly sought to keep the bombs from falling. It was not enough; and because it was not enough, and because time was needed to field its jet fighters and other wonder weapons, one final aerial gamble, was conceived.

They came in from the east, just above the trees, with the sun rising over their tails. Some two-dozen long-nosed Focke-Wulfs and a handful of sharklike Bf 109 fighters, their roaring engines abruptly shocking the British and Americans on the ground. Anti-aircraft artillery (AAA) crews scrambled to their guns and everyone else in the open ran for cover as the first day of 1945 opened with a shocking, unexpected bang.

There is nothing like a surprise attack to cure a hangover: instantly.

But the Germans flying in Operation Bodenplatte (Baseplate), had not welcomed the New Year with parties. At their bases the pilots were not permitted to return to private quarters following the evening mission brief, and they all remained stone cold sober during the evening's muted festivities. Despite the mauling given his Reich by the U.S. military, Hitler still had nothing but contempt for the American "mongrels," as he called them. A chaotic nation, he said, always in turmoil. His pilots knew better, and the Führer did not have to fight the Americans. Since his combat pilots did, they had no intention of going into battle with hangovers against a dangerous foe.

That morning the Luftwaffe gave its fighter pilots an unexpectedly robust breakfast of eggs and bacon, with real bread and coffee. Others might view it as a last meal for the condemned, but fighter pilots, even young inexperienced ones, never think like that. They are always going to win, no matter what, and this mission was for the future of Germany. Not the Reich or the Nazis, but for Germany. These pilots had grown up in the turmoil following the Great War and survived desperate years of the fragmented, ineffective Weimar Republic. If Bodenplatte, and the corresponding armored offensive on the ground, could prevent a similar future for their families, then these men would fight and die for it. At the very least, they hoped it would make the Allies consider peace, on some terms, and it would buy time.

For the pilots of JG 11 the mission was rebriefed at 0630,

and by 0800 Major Specht rolled down the runway at Biblis with his flight of four Focke-Wulfs. Eight Pathfinder Ju 88s followed, and for the next twenty minutes some sixty fighters from the three component groups of JG 11 got airborne from their bases on the east side of the Rhine. Assembling over Zellhausen, the wing headed northwest for Frankfurt and Koblenz. From there, JG 11 was to cross the Belgian border north of Aachen to make a time-over-target of 0920 hours. Twelve other fighter and ground attack wings were also assembling over Germany before heading northwest into Belgium, France, and the Netherlands.

The surprise was complete in northeastern Belgium that Monday morning near Asch. Ten guns from two AA batteries opened up and fired nearly 800 rounds, but although two Messerschmitts and one Focke-Wulf tumbled out of the sky, it didn't prevent the others from dropping their 500 kg bombs and strafing anything that moved.

But it was the wrong airfield.

Jagdgeschwader (Fighter Wing) 11 was supposed to hit the airfield at Asch, three miles to the south, but spotted Ophoven, next to the little town of Opglabbeek, and about half of the fighters attacked there. Asch (labeled Y-29), Ophoven (Y-32), and over one hundred little airstrips had been carved out of the woods by the U.S. 820th Engineer Aviation Battalion. Both of these were 5,000 feet long and 120 feet wide, with a pierced steel planking (PSP) runway. They looked identical, especially from the air. It didn't matter really, as snow had fallen and identification from the air was difficult. Add to it that the Ju 88 Loste, or Pathfinder, assigned to the mission, had attacked an American anti-aircraft unit on the way into the target and been damaged. After it turned back, navigation became more difficult, and as the fighters approached the Belgian border they ran into a blanket of heavy ground fog.

JG 11's commander, Major Günther Specht, was a thirty-year-old, one-eyed Prussian, and one of the few veteran aces surviving until 1945.* He knew there was no village near Asch so he nudged the stick of his fighter left and brought the remaining attackers around to the south. The strike could have been salvaged somewhat *if* they had come in from the east as planned—this would have put the sun directly into the defending Americans' eyes and made the attackers nearly impossible to see, a tremendous edge in the first few seconds of combat—and if over half his pilots had not hit the wrong target because they had less than ten missions' worth of experience.† And if, as Major Specht turned south, he hadn't run straight into eight American fighters that had just taken off from Asch.

But he did.

Contrary to a normal New Year's Eve, the one of December 31, 1944, was very different, at least to those in the European Theater of Operations (ETO). Two weeks earlier, at 0530 on December 16, 1,500 German heavy guns opened up along an eighty-mile stretch of the line from Belgium to France and, as in 1914 and 1940, the Wehrmacht suddenly thrust through the Ardennes. Concealed by heavy snow, the northernmost Sixth

---

* Thirty-four credited victories on the Western Front including fifteen heavy bombers and sixteen fighters. On May 23, 1940, Specht shot down a Spitfire flown by Squadron Leader Roger Bushell. Eventually incarcerated in Poland, Bushell, who engineered the famous "Great Escape" from Stalag Luft III, was subsequently caught and executed.

† During the initial fighting for the Bulge, JG 11 alone lost over 140 aircraft with fifty-nine pilots killed or wounded.

Panzer Army began its western assault to capture the bridges over the Meuse River bridges then on, hopefully, to Antwerp. This would bisect the Allies, deny them the huge port, and allow German resupply from Scandinavia. It would also, Hitler hoped, force a negotiation for peace. There was no reason, in his mind, why they would not settle. Then, with the west secure, he could then turn east and decisively deal with the oncoming Red Army. Even if ultimately unsuccessful, the delay meant time to amass enough rockets and jets to challenge the Allies.

It was a pipe dream.

The fantasy was concocted by Hitler's unbalanced and schizophrenic mind, but one that still held absolute power in the Reich. Field Marshal Gerd von Rundstedt, commander of all German forces in the west, protested, as did many general officers, yet to no avail. So when the armored spearhead of Tiger, Panther, and older tanks of the 1st SS Panzer Division rumbled west below the Elsenborn Ridge, the Reich's last big gamble, Operation Wacht am Rhein (Watch on the Rhine), had irrevocably begun.* Safe from Allied air support due to the weather, Wehrmacht and Waffen SS units shattered the thinly held center section of the line, and the 5th Parachute Division made an astounding twelve-mile initial penetration. Hasso von Manteuffel's Fifth SS Panzer Army was the middle mass of the three-pronged assault, and guarding the far southern flank was Erich Brandenberger's Seventh Army.

Speed was key.

---

* The code name was chosen to cover the buildup of troops in the border area. "Watch on the Rhine" was meant to look like a plan to defend the German border.

If the Meuse could be crossed in force before the weather lifted, before the Americans could reinforce, and before the panzers ran out of fuel, then the tactical portion of the plan just might succeed. The hammer stroke initially fell the hardest at Elsenborn Ridge on the eastern fringe of the Ardennes just a few miles from the German border. This ridge was the high ground parallel to a road that connected the main arteries leading into Liege and Antwerp. American field artillery, logistics personnel, and elements of the U.S. First and Second Divisions were emplaced at Camp Elsenborn. The line infantry units were under strenghth and exhausted after attacking across Belgium, and their replacements were green, untested troops. Nevertheless, they were dug in with plenty of ammunition, had relatively good communications, and were supported by artillery.

So stubborn was their defense that Kampfgruppe (Combat Gruppe) Peiper bypassed Elsenborn, leaving it to the 12th SS Panzer Division and 277th Volksgrenadiers—who failed to dislodge them.* It was the same, albeit on a smaller scale, all down the ridge. Pockets of American soldiers from various units, along with rear echelon clerks, cooks, and anyone else who could hold a rifle, resisted. This allowed defensive positions to be strengthened and rapid reinforcements occurred. Once the weather lifted, this included three squadrons of P-51 Mustangs from the Eighth Air Force's 352nd Fighter Group, which moved into Chièvres, Belgium. A detachment was for-

---

* Led by Lieutenant Colonel Joachim Peiper, infamous for the massacre at Malmedy of 150 captured American soldiers from the 285th Field Artillery Observation Battalion.

ward deployed to Asch, arriving on December 23, 1944, to sup-
plement the 366th Fighter Group's P-47 Thunderbolts.* Chuck
Yeager and the rest of the 357th Fighter Group were also back
in action. There had been a massive Eighth Air Force counterattack, some 2,000 bombers and over 800 fighters, deep into
the enemy rear. Harassment from the fighters and determined
resistance from ground units cost the Germans precious time
they did not have to lose.

There were other defensive positions that would not yield. St.
Vith was hurriedly reinforced with U.S. 7th Armored units and
the 82nd Airborne Division was thrown in to hold Cheneux,
just north of town, which blocked Kampfgruppe Peiper. Paratroopers, as always, were lightly armed with no heavy weapons
and limited supplies, yet the Bulge held out until December 21
against repeated assaults from the 1st Leibstandarte and 2nd
Das Reich SS Panzer Divisions. Suffering 80 percent casualties, the 82nd was pulled back, but the damage had been done
and the German timetable was wrecked.

Bastogne, to the southwest of St. Vith, had been immediately reinforced by a tank destroyer battalion and the 101st Airborne Division. By December 20 the town, a critical junction
with eleven hard-surfaced roads linking Liege and Antwerp,
was surrounded by elements of von Manteuffel's XLVII Panzer Corps, including the Panzer-Lehr and 26th Volksgrenadier
Divisions. Despite the grim tactical situation, Allied commanders realized that the Germans could not bypass Bastogne, and

---

* 328th, 486th, and 487th Fighter Squadrons: the Blue Nosed Bastards of
Bodney. The 366th FG had deployed to Asch from Laon-Couvron in November 1944.

with clearing weather they could be caught advancing on open ground. If this happened, they'd either have to retreat or dig in.

Like rocks in a stream, the American defenders forced the Germans to go around and disperse. This cost them momentum, further confused an already confusing tactical situation, consuming precious fuel and time, and irretrievably stalled the advance. On Christmas Eve, Hasso von Manteuffel suggested a withdrawal back into Germany, but Hitler refused point-blank. The day after Christmas four Sherman tanks, what remained from Company C, 37th Tank Battalion of Patton's Third Army, broke through to Bastogne. Just before 1700 the company commander, First Lieutenant Charles Boggess Jr., shook hands with a lieutenant from the 326th Airborne Engineers, 101st Airborne. Patton's armor continued up from the south and drove an armored wedge into the Germans. Along the Meuse River the British XXX Corps firmly held the bridges at Givet, Dinant, and Namur. Out of fuel and stretched thin, the Germans could either halt and hold their considerable salient or retreat eastward. They held.

Their last great hope was Operation Bodenplatte.

Unable to attack in conjunction with Watch on the Rhine due to weather, the air assault was now intended to cripple Allied airpower over the Low Countries so the ground offensive could continue. On December 31 Hitler also launched Operation Nordwind (North Wind) into the Alsace region 100 miles south of the Bulge along the French border. The idea was to split the French First Army and U.S. Seventh Army, then capture Strasbourg and its huge supply depots. This would permit a huge thrust up from the south to hook up with German forces in the Bulge. Hitler believed this would force the Allies to fall back and again be compelled to negotiate a peace. For the German combat generals this was nonsense, yet they had reached that desperate point in a fight where nothing remains except the fight, and the faint chance that in war anything could happen.

For Hitler, it was life or death for himself and his Reich, and he was perfectly willing to take as many down with him if the war was lost. For Eisenhower and the Allies, Operation Bodenplatte was a cold slap of reality in the face. They had underestimated the Germans and considered the war already won; it was just a matter of mopping up and sorting out the mess with the Soviets. For those in actual combat on both sides it was, as always, a day-to-day struggle for food, ammunition, a warm coat, and survival. The fighter pilots at Asch had it slightly better than their infantry brothers—but not much. They slept in torn tents over mud, had one small bucket of coal per day to keep warm, and slept in every piece of clothing they had. There was ammunition and fuel, but they were short of just about everything else.*

Yet the fighters kept flying.

About forty minutes prior to the German attack on Asch, eight P-47s from the 391st Fighter Squadron took off and turned south toward the heart of the Bulge. Led by Captain Eber Eugene Simpson, they caught an armored column in Ondenval, four miles southeast of Malmedy, and destroyed five Panzer Mark IVs. They then did what fighter pilots like best: went trolling for targets.† Back at Asch, the Mustang ground crews had

---

* Lieutenant John Stearns of the 486th Fighter Squadron flew with the one roll of toilet paper he'd brought with him.

† Captain Simpson, West Point '43, survived the war with 102 combat missions, a Silver star, and Distinguished Flying Cross, among others. One of

been up for hours preparing the planes for the morning mission while the pilots briefed. At 0915 a second flight of P-47s from the 390th Fighter Squadron rolled off the steel matting toward the southwest. Coming 180 degrees around to the northeast to join up their flight lead, Captain Lowell Smith saw anti-aircraft bursts over Ophoven and turned north to get a better look.

And he did—of at least thirty Messerschmitts and Focke-Wulfs heading south toward Asch and him. Radio calls filled headsets and gunfire arced across the sky as the airfield's two anti-aircraft guns opened up. Twelve 487th FS P-51s in three flights of four began their takeoff roll. Twenty-five-year-old Lieutenant Colonel John Meyer, their flight leader, was pulling his nose up and firing before his landing gear retracted, down-ing his first Fw 190 of the day right over the field.* For the next minute pilots scrambled to stay with their leaders, switch on gunsights, and jettison bombs or extra fuel tanks.

For the Germans it was worse.

Usually an attacker has the edge, but the confusion of hit-ting the wrong field threw the plan off. That, and running into twenty angry American fighters full of gas and bullets protect-ing their home field, such as it was. The P-47s split up the ini-tial attack long enough for the P-51s to get airborne, and the German pilots never recovered. Flights were split up; inexperi-enced wingmen flew through anti-aircraft fire, jabbered on the

---

the first P-80 jet pilots, in 1946 he was killed in an aircraft crash while going home on leave.

* J. C. Meyer would survive the war with 200 combat missions and 24 aerial kills. After graduating from Dartmouth, he would take the 4th Fighter Wing to Korea, lead the first all-jet dogfight in December 1950, and command SAC for Operation Linebacker during the waning days of the Vietnam War.

radio, and strafed the wrong targets. A derelict B-17 attracted nine or ten of them, and they consequently ignored rows of parked Thunderbolts and Mustangs. The runway wasn't hit, nor was the fuel storage area or ammunition dump attacked.

In the end, the thirty-minute battle at Ophaven and Asch cost JG 11 at least twenty-six fighters in air-to-air combat, confirmed by gun camera film or witnesses. Battery B and D of the 784th AAA AW Battalion claimed four and three kills, respectively, but this should be taken lightly. Living up to their unofficial motto of "if it flies it dies," anti-aircraft units fired more or less indiscriminately at whatever came close to them. Some 5 percent of the Luftwaffe's losses were from their own flak batteries who, for secrecy reasons, were not informed of the operation. The Allied units have no excuse, and pilots like Lieutenant Dean Huston of the 487th Fighter Squadron were lucky to have survived being shot down by friendly ground fire. Major George Preddy, commanding officer of the 328th Fighter Squadron, wasn't so fortunate. A P-40E Warhawk veteran from the Pacific, Preddy was now the leading P-51 ace in the ETO with 26.83 kills, including six Bf 109s on a single day back in August. He was shot down and killed by an American anti-aircraft unit on Christmas Day 1944: six days before Bodenplatte.*

Damage at Asch was light. The abandoned B-17 took most of the hits, though several aircraft were damaged and four enlisted men were injured. Ophoven fared worse; several men were wounded, and ten Spitfires were damaged or destroyed.

---

* This was likely a unit of the 430th AA Battalion. Preddy's brother Bill was a pilot with the 352nd Group at Asch. He would also be shot down and killed by AAA in early April 1945.

Across the ETO, results were similar. Tents, AAA sites, railroad equipment, assorted vehicles, and, by averaging out several sources, about 300 aircraft were destroyed with some 190 damaged. Only fifteen Allied aircraft were lost to aerial combat, with a dozen more damaged, including three P-47s and a Spitfire at Asch. Yet the material losses could, and were, replaced within days. Loss of trained pilots is always a severe blow as they cannot be manufactured like aircraft or weapons yet the United States, with its vast pool of resources, was not professionally hampered by the losses.

However, the numbers add up—and the truth usually lies somewhere in between—there is no doubt Operation Bodenplatte was a disaster for the Germans. Luftwaffe losses were *at least* 280 Fw 190s, Me 109s, and Ju 88s confirmed destroyed, with another 69 damaged. Out of approximately 850 aircraft, this amounted to 41 percent of the total lost.* Unsustainable, to be sure, but not as disastrous as the loss of 143 single-engine fighter pilots killed or missing (presumed dead), 70 captured, and 21 wounded. Forty-five *experten* were gone, with the cataclysmic loss of three wing commanders, five group commanders, and fifteen squadron commanders. These were all highly experienced veterans from multiple combat campaigns, and certainly after Bodenplatte the German air force effectively ceased to exist.

Historian Gerhard Weinberg accurately wrote that the Luftwaffe was "weaker than ever and incapable of mounting any

---

* Forty-seven percent from AAA; 23 percent from Allied fighters; and the rest from accidents, mechanical failures, and fuel.

major attack again." Truly, the duel for air superiority was unquestionably finished in the skies above Europe, and even some
sort of air parity on any level was now impossible—time had
run out. So by the desperate spring of 1945, after Bodenplatte,
all that remained for Germany, and the Luftwaffe in particular,
were desperate hopes like the jet fighter, and even more desperate measures like the rocket-powered Natter.

But time had caught up with Lothar Sieber as it had for hundreds of other Luftwaffe pilots. Hans Mutke, two weeks after the dive, flew his Me 262 across the border into northern
Switzerland and landed at an airfield outside Zurich. Claiming
he was lost, Mutke was interned for the short duration of the
war. His unit, 9 Staffel of Jagdgeschwader 7, fought the last
Second World War European dogfight thirteen days later on
May 8, 1945.

Near the time Mutke made his dive, Bob Hoover escaped
from the notorious Stalag Luft I in northern Germany.* He
passed through Russian lines and, thanks to Gus Lundquist's
instruction, managed to steal a Focke-Wulf 190 from an abandoned German airfield. Getting airborne and very aware he
was flying an enemy fighter into U.S.-held airspace, Hoover
stayed up long enough to get to Holland, then crash-landed in
a farmer's field. Over the tips of their pitchforks he barely con-

---

* Gabby Gabreski, Doolittle Raider Charles Greening, and 479th Fighter
Group commander "Hub" Zemke were also prisoners here. So was future
actor Donald Pleasence who, ironically, starred in the 1963 movie *The Great
Escape*, about Stalag Luft, III.

vinced the local farmers to take him to a British unit nearby rather than perforate him. For the first time in sixteen months Bob Hoover was a free man.

Ken Chilstrom had survived eighty combat missions in the A-36 and was finally back in the States. One day in late November 1943, as he and another pilot, Gene Santella, were sitting in their four-man tent in Italy, K.O. was surprised by the 27th Fighter-Bomber Group commander, Colonel Dorr Newton. "He came and sat down," Chilstrom recalls, "and said he was here because the morale of the outfit was going downhill. He was right. There was supposedly a rule, which wasn't really enforced, that with fifty missions in fighters you rotated home. Well, I had eighty missions and my tent mate had eighty-seven . . . as far as I knew no one had ever left unless they'd crashed and burned."

The colonel asked them if they wanted to go home and "it didn't take too long to decide that," Chilstrom remembered. The two officers found themselves with orders back to "some damn place in Florida" but no official travel arrangements. "You cut loose a pair of young lieutenants who want to get home and they'll find a way!" Indeed they did. Chilstrom and Santella hitchhiked down to Naples and caught a flight back to Casablanca. From there they scrounged a ride on a B-24 returning to the States through Val de Cans outside Belém, Brazil. After that it was the Redistribution Center at Camp Miami Beach in Florida. Over 300 apartments and hotels had been requisitioned by the Army, and it was here that combat vets were debriefed, given new uniforms and physicals, then released or reassigned.

After a week or so of this, Chilstrom was asked what he wanted to do and he said he wanted to go to Wright Field. "They tried to send me to a P-40 replacement training unit in Way-cross, Georgia," Ken recalls, "but I wanted nothing to do with training. I knew about Wright Field and thought that's where I oughta go." So he went to see the personnel officer and said, "these orders aren't at all to my liking . . . I've got a suggestion. I'd like to have you send me to Wright, and I'll find a job when I get there." This was not the way things were done, but for some reason the personnel officer "went along with it, cut orders and I rode the train up from West Palm Beach to Dayton, Ohio."

Chilstrom found the correct building, got up to the second floor, and found the Flight Test Division's executive officer, Lieutenant Colonel Ernest K. Warburton, who listened to his story and sent the young pilot along to Major Chris Petrie, the Chief of Fighter Test. Petrie told him there was no need for more pilots as he just got a few men in from the South Pacific, but "we could use an officer in maintenance." Ken immediately told the major how he'd been trained as a maintenance techni-cian at Chanute and, stretching the truth a bit for a good cause, conveyed the idea that he could easily handle such a position. He was hired on the spot and found himself as the assistant maintenance officer for the Fighter Test Section at Wright Field.

George Welch was also home.

After nearly eighteen months in the Pacific flying P-39 Airacobras and P-38H Lightnings, he had shot down at least sixteen Japanese aircraft over the course of 348 combat mis-sions. Following his marriage to Janet, he did in fact come home about the same time Ken Chilstrom left Italy. George went to Florida into a tactics development unit and reluctantly resumed publicity work for the Army. Early in 1944 he had a way out; the chief engineering test pilot for North American

Aviation, Ed Virgin, offered him a job using his combat experience to flight test variants of the P-51 Mustang. George immediately accepted a position as a test pilot with North American Aviation during the spring of 1944.

Chuck Yeager left the European Theater in January 1945, immediately after Bodenplatte and the Battle of the Bulge. With the combat debut of the Me 262 and the disappointing performance of the Airacomet, Lockheed's Shooting Star had been approved for accelerated service testing. The result was a little-known historical footnote called Project Extraversion. On December 30, four days after Bastogne was relieved and the day before Hitler's big push into Alsace, a freighter slipped into the river Mersey on England's northwest coast. A collection of unmarked crates were offloaded, then taken from Liverpool to nearby RAF Burtonwood: the first two American jet fighters had arrived in Europe.

Colonel Marcus Cooper and Major Fred Borsodi of Wright Field were to fly a pair of YP-80s for operational testing and evaluation. Two other jets appeared at the Foggia Airfield Complex in southern Italy during January and, in both cases, the aircraft were entirely off limits to regular British and American military personnel. Veiled in secrecy, these aircraft were maintained by civilian contractors but flown by Army pilots. For the pair in Britain, this was an extremely short operation. With one of the jets reassembled by the end of January, Colonel Cooper flew the first test hop without incident. Unfortunately, the next day, January 28, 1945, hot exhaust gases vented directly into the YP-80's tail section and it disintegrated. The jet crashed into a field, killing Major Fred Borsodi, Yale graduate and 130 combat mission veteran.*

The two YP-80s in Italy flew out of Lesina, near the Adriatic coast, on what may have been armed reconnaissance missions and, technically, combat. Very few people were even aware of the jet's existence, yet the Shooting Stars did fly, and Major Ed LaClare logged a pair of operational sorties to an unspecified location north of the base. Both aircraft were returned to the United States, while the remaining YP-80 in England was loaned to the Royal Aircraft Factory, Farnborough, to test the new Rolls-Royce Nene turbojet.

Despite the Luftwaffe's success with the Me 262, the Komet, and the Arado 234 jet bomber, after the Bodenplatte disaster, time quickly ran out for Germany. Bases were abandoned, files burned, aircraft destroyed, and those who could flee westward did so to escape the advancing Russians. Tragically, President Franklin Roosevelt did not live to see the culmination of his extraordinary efforts to extricate the United States from the Great Depression, and consequently deliver civilization from the darkness of imperialism, fascism, and national socialism. On April 12, 1945, the man who had done so much to shape his world (and ours) died at his "Little White House" at Warm Springs, in far western Georgia. That same day Harry S. Truman was sworn in as the thirty-third president of the United States, and when he asked Eleanor Roosevelt if he could help in any way, she replied, "Is there anything we can do for you? For you are the one in trouble now!"

---

* YP-80 # 44-86026.

In January 1945, the Red Army had crossed the Oder River and closed to within forty miles of Berlin. Over seven million artillery shells had been stockpiled and 300 artillery pieces allocated for every square mile of the German capital. Several thousand ground-attack aircraft had been moved forward as well; Stalin wanted his revenge, and eight American and British armies were ordered to halt at the Elbe River so he could have it. Fragmented Wehrmacht battalions banded together with a few Waffen SS and Panzergrenadiers units who knew there was no mercy for them, and they prepared to fight to the death.*

On April 20, Russian armor overwhelmed Hasso von Manteuffel's few remaining tanks and turned toward Berlin. By April 26, while Hitler cowered in his bunker and expelled Göring and Himmler from the Nazi Party, a half-million Red Army soldiers began looting the city. At the same time in San Francisco, representatives of fifty countries gathered to produce a charter that would structure a better postwar world, a world of international cooperation overseen by what they called the United Nations. This was envisioned as a peaceful embodiment of Roosevelt's Four Freedoms, including Freedom from Fear, which still had to be won. At 1530 on April 30, as the Soviet 8th Guards Army leveled downtown Berlin, the leader of Germany's Thousand-Year Third Reich married his mistress, shot his dog, and finally killed himself; Victory in Europe (VE) Day was officially designated a week later on May 8, 1945.

---

* This included Swiss, Danish, Spanish, and Swedish volunteers of the 11th SS Panzergrenadiers. The SS 33rd (Charlemagne) Division, an all-French unit, was also present.

Three months after this, nearly 7,000 miles to the east, a single B-29 under the call sign of "Dimples 82" took off from the Pacific island of Tinian. At 0815 local time in Japan, a ten-foot-long, oddly shaped bomb named "Little Boy" detonated 1,968 feet above the city of Hiroshima.* Surviving locals later remembered it as a "sheet of the sun." Three days later, on August 9, another Superfortress dropped a second atomic bomb on the port city of Nagasaki. President Harry Truman, himself a combat veteran of World War I, knew this was the only way to defeat the Japanese, and he was correct.† After declaring that the empire would never know defeat, and absolutely intending to cause millions of American casualties, the Japanese unconditionally surrendered on September 2, 1945, under the threat of additional such bombings.‡

The Second World War, the most destructive cataclysm in human history, was over, and with it the old world order that had guided humankind for centuries faded also. Men emerged from the firestorm who, for better or worse, had been shaped by the economic collapse of 1929, the bleakness of the Great Depression, and years of war. Many of them could never return to a normal day job, or a school, or any other fragment of their

---

* The first B-29 was named *Enola Gay* and flown by Colonel Paul W. Tibbets. The second B-29 was called *Bockscar*, commanded by Major Charles W. Sweeney.

† Truman was an artillery officer with the 129th Field Artillery and fought during several 1918 engagements, including the Meuse-Argonne. He also volunteered to fight in World War II, and remained an inactive reserve officer until 1953.

‡ The Joint Chiefs of Staff estimate that 1.2 million American casualties, and at least 5 million Japanese, were spared when invasion became unnecessary due to the employment of the atomic bomb.

old life. They were no longer suited to a world without danger, excitement, and the constant challenge of life over death.

The next great age had arisen, an age of polarizing ideologies, to be sure, where the world's fate often hung by a nuclear thread and the awesome, destructive power of weapons conceived during World War II would dictate the pattern of life for the next five decades. Yet through this, humankind managed to come to grips with its own ability to change the world. Men knew now, as never before, that limitations were barriers that *could* be swept away by those with the brains, guts, and courage to do so.

Part Three

HUNTING THE DEMON

Flinders and flames, flinders and flames,
The names soon forgotten, along with the blames
God bless Pratt and Whitney and Wrights'
And pray for doomed ghosts in aluminum kites.

WRIGHT FIELD: THE READY ROOM

ANONYMOUS

Seven

World War to Cold War

By the summer and fall of 1945, strange captured German aircraft appeared in the sky over the Midwest United States, and Ken Chilstrom, now in the Fighter Operations Section of the Flight Test Division at Wright Field, was thrilled. "Everyone knew the jet was the future," he recalls. "The rocket stuff was interesting in that it was the fastest way to blow through the so-called sound barrier, but it had no practical applicability to tactical operations." Vast amounts of technical data had been discovered all over the crumbling Reich, and much of it made its way west to Britain and the United States.

In fact, many programs and operations were put into play toward the end of the war precisely to collect advanced technology, and the scientists who created it. One such operation was Lusty (**LU**ftwaffe **S**ecret **T**echnolog**Y**), a U.S. Army Air Force initiative organized by the Exploitation Division at Wright Field, where Ken Chilstrom, Glen Edwards, Gus Lundquist, and other first-rate test pilots were now stationed. Lusty was

divided into two main teams; Team Two was to collect documents and round up any scientists while Team One, working
from top-secret "black lists," scooped up any aircraft and weapons they could find.

Colonel Harold Watson, chief of intelligence (T-2) at Wright
Field, led Team One into Germany as the war ended. A flam-
boyant former test pilot with a master's degree in aeronautical
engineering, he further divided his team into two branches,
one to pick up rockets and piston-engined aircraft while the
other branch went after any jets. Watson's Whizzers, as they
were called, acquired a veritable treasure trove of Luftwaffe
secrets. These included an Arado 234, the world's first high-
altitude reconnaissance aircraft and jet bomber; a Dornier 335;
some Me 262s; the Me 163 Komet, and even a rare surviving
Natter.* Ten Me 262s were flown across Germany and through
France to Querqueville Airfield outside Cherbourg, where they
were loaded onto HMS *Reaper*, an escort carrier, and shipped to
the United States.

In the meantime, Ken Chilstrom had happily spent his first
six months home as the maintenance officer for the Flight Test
Division. In this position he got to fly every aircraft that had
been in for maintenance, building up his flying time and im-
pressing Major Chris Petrie, the chief of the Fighter Test Section, with his skill and quiet, unassuming demeanor. When
there was an opening in the Performance Section, Ken was
able to slide right in. It was here that he met Colonel Harold
Watson and began flying some of the exotic captured aircraft

---

* Watson was rumored to be the inspiration behind Steve Canyon, Milton
Caniff's comic book hero.

the Whizzers and others had brought back. These included the Me 262, an A6M "Zero," the XP-59 Airacomet, and Chilstrom's favorite, the Focke-Wulf Fw 190.

"It was all stopwatches and knee cards," Ken recollects. "Nothing fancy and we had no formal training as test pilots because there was no such thing in 1944." This was a situation soon to be remedied, and in September the Engineering Flight Test School was established at nearby Vandalia airport, graduating the inaugural class in 1945. Ken's classmates included Dick Bong, America's highest-scoring ace; Tony LeVier; Fred Ascani; and Glen Edwards, with whom he shared a house.

Edwards had graduated from UC Berkeley with a degree in chemical engineering and gotten into the Army Air Corps five months before Pearl Harbor. Like Ken Chilstrom, Edwards fought in North Africa, Tunisia, and Sicily, though the Canadian native flew the A-20 Havoc light bomber. Coming home with four Distinguished Flying Crosses in December 1943, Edwards would meet up with Chilstrom in 1944 at Wright Field. "Glen was my best friend," Ken recalls wistfully. "He was a wonderful pilot and all-around great guy."

It became apparent to everyone that the advanced German aviation programs were far ahead of the Allies in terms of aerodynamics; there were swept wings, including a *forward-swept wing*, vertical tails, and a host of other innovative applications backed by stacks of wind tunnel data and practical flying test experience. The Germans were incorporating bubble canopies, ejection seats, and pressurized cockpits into their aircraft. Though scores of designs were captured, two in particular bear discussion for the impact they had on postwar development and the quest for supersonic flight.

Messerschmitt's P.1101 arose from the 1944 Emergency

Fighter Program and was designed around the smallest, lightest airframe possible paired with the most powerful engine available: the Heinkel HeS 011 turbojet. With its tricycle landing gear and high tailplane, the P.1101 was intended as a sleek, deadly successor to the Me 262. The swept, shoulder-mounted wings were ground adjustable between 35 and 45 degrees, which the Germans considered vital for raising the critical Mach number and reducing the effects of transonic shock waves—another breakthrough as yet unrealized in America or Britain. With its blunt, nose-mounted intake and bubble canopy the Messerschmitt looked like a stouter, older brother to North American's premier jet fighter: the North American Sabre. In fact, a nearly completed P.1101 prototype was discovered at Messerschmitt's project facility in Oberammergau, and much of the German data would be used to refine the Sabre's design.

The other aircraft of note was Focke-Wulf's Ta 183. Designed by Hans Multhopp as a single-seat, single-engine fighter, the 183 was intended, from its inception, to operate in the transonic region. Multhopp calculated that this not only necessitated a pronounced sweep, but also very thin wings that would focus the shock wave outboard away from the aircraft. Control would be more stable this way and was further enhanced by a high tailplane that kept the rear surfaces above the burbling transonic wake. The vertical fin was swept back 60 degrees and topped by a dihedral (upward angle) tailplane.* It also incorporated *el-*

---

* Initially described by Sir George Cayley in 1810, a dihedral in aerodynamics denotes an upward (positive) angle between two horizontal surfaces. The upward angle is a stabilizing influence against an aircraft's rolling tendencies.

evons, a relatively large control surface on the wing's trailing edge that combined the functions of elevator and aileron.

Stubbier than the P.1101, the 183 was heavier and faster with a top design speed of 593 miles per hour. Its thin wings could not support interior armament so, like the Messerschmitt (and the Sabre), cannons were to be nose mounted on either side of the intake. One can see shades of the future Soviet MiG-15 in its design, and this is hardly surprising as the Russians also took whatever advanced technology they could, including most of the V-2 rocket program. Given the scope of available technology, German ingeniousness was futuristic and a decade ahead of its time. No less impressive was that even faced with constant bombing, a chronic shortage of materials, and, for those realistic enough to admit it, a lost war, these various programs endured until the very end.

But German limitations, namely their jet engine manufacturing issues, were as staggering as their successes and in this area the Allies were well ahead. Engines like the Rolls-Royce Derwent V, basically a powerful, polished derivative of Frank Whittle's W.2B, used Nimonic 80, a high-temperature resistant alloy, for its turbine blades and mated the engine with Gloster's Meteor. Rolls-Royce also utilized the Lockheed YP-80 as a test bed for its improved Nene engine, then sold it under license to Pratt & Whitney for use in the U.S. Navy's Grumman F9F Panther.

Britain, however, under a new government, was determined to cut budgets and fell rapidly behind after the war. The Miles M.52 supersonic program was scrapped and the Americans, specifically Bell Aircraft, were able to use the British data to complete its X-1 rocket plane. What truly spurred U.S. development from 1945 onward was the synthesis of German aerodynamics with initial British test data and American government financial backing. The impact of German technology and design would be immediately felt within both Eastern and Western aerodynamic spheres. Interestingly, some innovations, like the swept wing, were independently discovered, albeit a decade late.

American Bob Jones of the NACA had paralleled German research and, in early 1945, was convinced he had discovered the solution to tame transonic airflow: the swept wing. That summer a team of American engineers under the auspices of Operation Paperclip rounded up every German scientist and technician possible. One group, which included Hugh Dryden and Boeing's George Schairer, was led by Theodore von Kármán himself, who, due to his fluency in German, education in Germany, and personal contacts with many scientists, was an immense asset. Reunited with Adolf Busemann, whom he had not seen since the 1935 Volta Conference, the Hungarian asked, "What is all this about wing sweep?" and Busemann's face lit up as he replied, "Oh, you remember, I read a paper on it at the Volta Congress in 1935."

In fact, no one did remember until Busemann's original work surfaced and that, along with the German's hard, practical data, expedited his transfer from the British sphere to the Americans. He, like Alexander Lippisch and hundreds of others, found a new life in the United States while those much less fortunate ended up in Soviet hands. In the end, both sides would rapidly overtake the German lead, yet it cannot be denied that the war, and subsequent Allied victory, opened the door into a new age.

I t was an age, almost immediately, of great promise and hope mixed with tremendous suspicions and fears. The war was

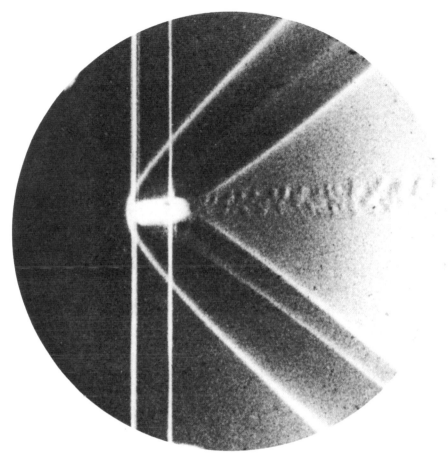

Austrian physicist Ernst Mach's astounding 1887 schlieren photograph that clearly reveals shockwaves forming on the bow of a supersonic projectile.

The author with one of two surviving Ba 349 "Natters." This one is stored in the National Air and Space Museum's Paul E. Garber Facility in Suitland, Maryland. *(Courtesy of the author)*

Luftwaffe test pilot Lothar Sieber. There are unproven rumors that Sieber exceeded Mach One in a Natter on March 1, 1945, as he plummeted to his death.

The Bell XP-59A Airacomet, America's first jet fighter, in flight over Rogers Dry Lake during the fall of 1942. *(Bell Aircraft archives)*

Captain George Welch in his P-38 Lightning, at war in the Pacific. Welch survived 348 combat missions and was credited with 16 confirmed aerial kills. Welch's WWII service is also notable for the fact that he was one of the very few American pilots to get airborne on December 7, 1941, during the attack on Pearl Harbor; credited with four kills on that day, Welch received the war's first Distinguished Flying Cross, a distinction he shared with his comrade Kenneth Taylor. *(National Museum of the U.S. Air Force)*

Flight Officer Chuck Yeager with the original "Glamorous Glen," a P-51B, in early 1944. Yeager would fly 64 combat missions and end the war with 12.5 confirmed aerial victories. *(National Museum of the U.S. Air Force)*

North American A-36 Apache. North Africa, 1942. Ken Chilstrom flew this air-to-ground version of the P-51 Mustang during 80 combat missions over Sicily and Italy. *(National Museum of the U.S. Air Force)*

XS-1 under construction at Bell Aircraft's factory, 1945.

The XS-1 offered a unique opportunity for transonic aerodynamic research. Shown here is a wide and varied array of recording and research packages displayed across nearly the entire wing. *(NASA)*

Bell X-1A with military markings after the USAAF took over the project in June 1947. Yeager is first on the right near the cockpit hatch. *(Guy Aceto Collection)*

The second X-1 of three built by Bell. The black Xs on the fuselage are utilized for photo calibration. *(Guy Aceto Collection)*

South Base of Muroc Army Air Field next to Rogers Dry Lake. The regular army (and later air force) used this area while the secret programs were conducted on the much smaller North Base. *(Guy Aceto Collection)*

LEFT: What passed for an Officer's Club—Muroc Army Air Field, circa 1945. The military gave little thought to comfort or entertainment, which was one reason why Pancho's Happy Bottom Riding Club was so popular. *(Guy Aceto Collection)*
RIGHT: The Happy Bottom Riding Club was the focal point of social life around Muroc in the late 1940s. The pilots loved it, and their wives hated it—there were no secrets here about any of the military classified programs. *(Guy Aceto Collection)*

The Happy Bottom hostesses were obvious attractions for men forced to live in spartan conditions and very often separated from their families. Pancho Barnes is seated in the middle with the dog on her lap. *(Guy Aceto Collection)*

Bell test pilot Jack Woolams. Wearing a bowler hat and gorilla mask, he would often sneak up on conventional military aircraft in the XP-59 jet. The first pilot of the XS-1, he was killed in August 1946 while preparing for the first postwar National Air Races. *(Guy Aceto Collection)*

Jack Woolams and the XS-1. He was the only pilot to fly the rocket plane over Florida, and made its first powered flight on December 9, 1946. *(Bell Aircraft archives)*

Chalmers "Slick" Goodlin took over as Bell's chief test pilot after Woolam's death. Eager to fight, Goodlin became the youngest commissioned officer in the Royal Canadian Air Force and flew Spitfires in England until late 1942. Returning to the U.S., he became a Navy test pilot before joining Bell, and made 26 flights in the XS-1 before the program was turned over to the USAAF. *(Bell Aircraft archives)*

The racing version of the Shooting Star, the P-80R, roaring past a speed marking station at Muroc Army Air Field. *(National Museum of the U.S. Air Force)*

Colonel Al Boyd, June 1947, after setting a world speed record of 623.7 mph in a P-80R Shooting Star, returning the honor to the U.S. after 24 years. Rightfully called the father of the U.S. Air Force test pilot program, and as chief of the Flight Test Division at Wright Field, Boyd was Chilstrom and Yeager's commander. *(National Museum of the U.S. Air Force)*

Major Ken "K.O." Chilstrom after delivering the first piece of mail by jet aircraft in June 1946 to Orville Wright. Ken was chief of the Fighter Operations Section and the Flight Test Division's first choice to fly the XS-1, but he declined the offer to run the USAF YP-86 Sabre test program. *(Colonel John Chilstrom)*

George Welch, with his distinctive red flight helmet, flying the North American XP-86 Sabre over the Mojave Desert. Destined to become the mainstay of USAF fighter aviation during the Korean War, the Sabre was quite capable of supersonic flight while diving, which Welch officially accomplished on November 13, 1947. It was an open secret around Muroc and Pancho's that he had flown past Mach One in the XP-86 on at least one occasion prior to the official USAF X-1 flight of October 14, 1947. *(Guy Aceto Collection)*

Three of a kind. North American's legendary XP-86, the enigmatic test pilot George Welch, and his classic MG sports car in Los Angeles, 1947. *(National Museum of the U.S. Air Force)*

With its four-minute fuel supply, the X-1 had to be air dropped at altitude in order to accelerate to Mach One. An early loading technique was to raise the carrier aircraft, a B-29, on jacks and secure the rocket plane to its belly.
(Guy Aceto Collection)

Jackie Ridley at Muroc. A first-rate test pilot himself, Ridley had the background in aerodynamics that Yeager lacked. Possessing a master's of aeronautical engineering from Caltech, Ridley personified what the new USAF sought in its postwar test pilots; a solid mix of experience and formal education. *(National Air and Space Museum)*

Piloted by Jack Woolams, the XS-1 prepares to drop over Pinecastle, Florida, in 1946. *(National Air and Space Museum)*

Like the fighters he'd flown in Europe, Chuck named X-1 #6062 after his wife, Glennis Dickhouse Yeager. *(NASA)*

X-1 cockpit. The Machmeter is second from the right on the top row. Note that there is no feasible way for a pilot to bail out once the door on the right side is locked. *(National Air and Space Museum)*

Photographed by Bob Hoover from a P-80 chase plane, the X-1 lights off its Reaction Motor rockets over the Muroc Test Range. *(NASA)*

Date: ~~18~~ 14 October 1947

Pilot: Capt. Charles E. Yeager

Time: 14 Minutes

9th Powered Flight

1. After normal pilot entry and the subsequent climb, the XS-1 was dropped from the B-29 at 20,000' and at 250 MPH IAS. This was slower than desired.

2. Immediately after drop, all four cylinders were turned on in rapid sequence, their operation stabilizing at the chamber and line pressures reported in the last flight. The ensuing climb was made at .85-.88 Mach₁, and, as usual, it was necessary to change the stabilizer setting to 2 degrees nose down from its pre-drop setting of 1 degree nose down. Two cylinders were turned off between 35,000' and 40,000', but speed had increased to .92 Mach₁ as the airplane was leveled off at 42,000'. Incidentally, during the slight push-over at this altitude, the lox line pressure dropped perhaps 40 psi and the resultant rich mixture caused the chamber pressures to decrease slightly. The effect was only momentary, occurring at .6 G's, and all pressures returned to normal at 1 G.

3. In anticipation of the decrease in elevator effectiveness at speeds above .93 Mach₁, longitudinal control by means of the stabilizer was tried during the climb at .83, .88, and .92 Mach₁. The stabilizer was moved in increments of 1/4 - 1/3 degree and proved to be very effective; also, no change in effectiveness was noticed at the different speeds.

4. At 42,000' in approximately level flight, a third cylinder was turned on. Acceleration was rapid and speed increased to .98 Mach₁. The needle of the machmeter fluctuated at this reading momentarily, then passed off the scale. Assuming that the off-scale reading remained linear, it is estimated that 1.05 Mach₁ was attained at this time. Approximately 30% of fuel and lox remained when this speed was reached and the motor was turned off.

5. While the usual light buffet and instability characteristics were encountered in the .88-90 Mach₁ range and elevator effectiveness was very greatly decreased at .94 Mach₁, stability about all three axes was good as speed increased and elevator effectiveness was regained above .97 Mach₁. As speed decreased after turning off the motor, the various phenomena occurred in reverse sequence at the usual speeds, and in addition, a slight longitudinal porpoising was noticed from .98-.96 Mach₁ which controllable by the elevators alone. Incidentally, the stabilizer setting was not changed from its 2 degrees nose down position after trial at .92 Mach₁.

6. After jettisoning the remaining fuel and lox a 1 G stall was performed at 45,000'. The flight was concluded by the subsequent glide and a normal landing on the lake bed.

CLASSIFICATION CHANGED TO Unclsf----
WADE TUX WCIPP-H-25 MM
AFB 29 Oct 57   By _____ Yeager
CHARLES E. YEAGER
Capt., Air Corps

Hamill   7 7/nr 58

"The needle of the machmeter fluctuated . . . then passed off the scale.": The original supersonic flight test report written by Captain Chuck Yeager on October 14, 1947. *(National Air and Space Museum)*

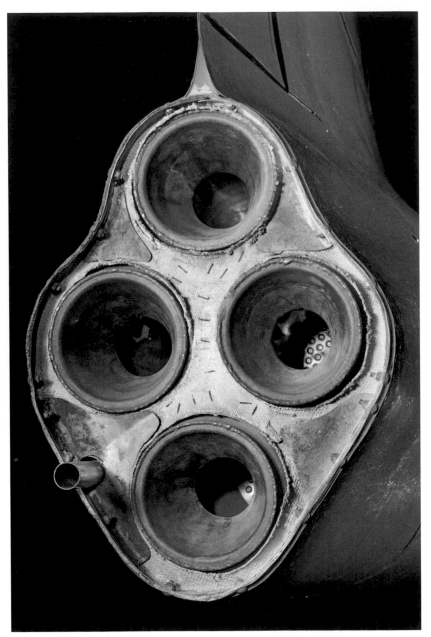

Reaction Motors XLR11 rocket engine. The four rockets were fueled by a mix of ethyl alcohol, water, and liquid oxygen to produce 5,900 pounds of thrust. This gave the X-1 roughly four short minutes of powered flight, thus necessitating an air launch procedure from the B-29 mother ship. *(National Air and Space Museum)*

George Welch over the Dry Lake in North American's YF-100A Super Sabre. Repeating his performance with the XP-86, Welch took the F-100 supersonic during its first flight on May 25, 1953 over Palmdale airport. *(Guy Aceto Collection)*

North American YF-100A with Welch in the cockpit. George would disregard stability warnings from USAF test pilots Chuck Yeager and Pete Everest regarding the small vertical tail, and perish on October 12, 1954, when his Super Sabre came apart in flight. *(Guy Aceto Collection)*

Northrop YB-49 "Flying Wing." This one was a sister ship to YB-49 #42-10238, which crashed on June 5, 1948, killing Major Glen Edwards. Edwards was Ken Chilstrom's best friend and also passed on the X-1 program in order to complete a master's of aeronautical engineering at Princeton. Muroc Air Force Base would be renamed Edwards AFB in his honor. *(National Museum of the U.S. Air Force)*

Bell X-1 #6062, first aircraft to officially pass the speed of sound, forever airborne in the National Air and Space Museum. *(National Air and Space Museum)*

over and the men who fought it came back, physically at least, but, like all combat veterans, each man mentally returned whenever he could. Some, like Ken Chilstrom and George Welch, were in combat early so they were able to begin picking up the pieces before the end of the war. Ed Virgin, chief of Engineering Flight Test and the chief test pilot at North American Aviation, had hired George Welch away from the Army, so by 1945 he and Jan were in Los Angeles with George deep into P-51 follow-on programs.

In Ohio, Ken Chilstrom was working directly under Gus Lundquist, who was liberated from Stalag Luft 1 during May 1945 and had taken over the Fighter Test Section in June. Lundquist, whom Ken considered "the finest test pilot I ever knew," was succeeded by Francis "Gabby" Gabreski, who had also survived Stalag Luft 1 and took over the section in September.

Chuck Yeager came home in early 1945 and went straight to California to propose to Glennis Dickhouse. They were married on February 26, at his childhood home in Hamlin, West Virginia.

Following a brief stint as an instructor pilot, Yeager arrived at Wright Field in July just weeks before the end of the war. He had chosen the base because Glennis was pregnant and Ohio was as close as he could get to West Virginia. One fall day while circling the field in an Airacomet, a P-38 appeared and a mock dogfight ensued. Neither could gain an advantage and Yeager called a "knock-it-off" to end the fight. Both landed. Yeager had been astounded to see a Lightning flown that way and went out of his way to meet the pilot. "Man . . . I didn't know the 38 could swap ends like that," he said as he shook the other officer's hand. Bob Hoover, ever pugnacious, snapped back with, "Those eyes of yours were bigger than a stripper's knockers."

Hoover had been through a particularly nasty time and was just happy to have survived the war. During his sixteen months as a POW, he had absorbed everything Gus Lundquist told him about enemy fighters. Hearing the Germans planned to murder the 10,000 prisoners in Stalag Luft 1, he escaped in April 1945, right before the war ended. Recaptured by the Russians, Hoover again escaped while his captors were trying to figure out how to use an indoor flush toilet, something none of them had ever seen.

So it came to be that during 1945 all these pilots, different in many ways but linked together by combat and flying, found themselves together at Wright Field. It was the age of the jet, and possibilities seemed endless. On October 24, 1945, within weeks of Japan's surrender, the United Nations was founded with the aim of avoiding another such world war. Forty-six nations were initially part of this effort, plus the five Allied powers who composed the Security Council: China, France, the Soviet Union, the United Kingdom, and the United States. The opening lines of the UN Charter expressed its lofty goals thus:

"We the peoples of the United Nations determined to save succeeding generations from the scourge of war, which twice in our lifetime has brought untold sorrow to mankind, and to reaffirm faith in fundamental human rights, in the dignity and worth of the human person, in the equal rights of men and women and of nations large and small . . ."

High-minded and admirable, to be sure, but most of the world, and particularly America, wanted to live a little and make up for time lost during the dark years. Solo artists, who traveled easier and far cheaper than the extravagant prewar big bands, were now the rage. Bing Crosby, Doris Day, Ella Fitzgerald, and, above everyone else, Frank Sinatra, came to symbolize the postwar era. At one concert over 10,000 fans lined Times Square, and the singer was pelted with bras and panties when he took the stage.

American boys sporting crew cuts and cuffed pants were hanging around drugstores or soda fountains looking, of course, for girls. The girls still favored sweaters and skirts, mostly of fairly drab material due to wartime shortages, but that was changing fast. Americans were undeniably becoming more casual, for a variety of reasons. As with most who have passed through a great upheaval, the survivors felt they deserved to relax a bit, and who could blame them? Men returning home were understandably tired of uniformity and wanted to loosen up. They wanted color, and they got it in a variety of suits and especially neckties. Jackets were often worn with no tie, or not at all, and men often favored the loose, untucked shirts they had seen in Hawaii or California. With no material rationing, pants were cut fuller with wide cuffs, and suit jackets much longer than during the war. Quickly discarding their military-style haircuts, men began wearing their hair longer with pomade or gel.

Girls kept the shortened sleeves and knee-length hems imposed by war shortages, but a variety of colorful fabrics were again available, and the bobby-soxer made her debut. Pleated skirts worn with socks rolled down over saddle shoes became a symbol of the "American look," as *Life* magazine called it. Older women wore hats of all kinds, accompanied with cotton gloves since they had not been rationed and could be dyed. Inspired by military uniforms, women wore shoulder pads with knee-length skirts, though pants, which they had worn by necessity in wartime factories, were becoming more common.

In general, women emerged from the Second World War as

their mothers had from the Great War, more independent and less willing to submit to what many believed were outdated societal demands. The marriage and birth rate had increased by 50 percent from prewar levels, though a surprising number of these unions were encouraged by "Allotment Annies"; women who married multiple soldiers to receive the extra $50 per month allotment or, if they were "fortunate," the $10,000 death benefit paid to combat widows. "Dear John" letters became more common as women, for different reasons, decided not to wait for soldier husbands, fiancés, or boyfriends, and at least 650,000 American children were born out of wedlock during the war years. Yet many American females were extremely patriotic, very proud of their country and their fighting men, and waited faithfully until the soldiers came home and life went on.

By 1945 over nineteen million women made up 36 percent of the civilian workforce—and they liked it. "They put me into a training program for about two months," one Connecticut woman recalled. "Oh, I was so pleased with myself; it was for the war effort." Understandably, many women were not pleased with the postwar mass firings that occurred so returning soldiers could have jobs. "I know the pride I had felt during the war," one woman later wrote. "I just felt 10 feet tall. Here I was doing an important job, and doing it well, and then all at once here comes V-J day [Victory in Japan] and I'm back making homemade bread."

At one point in 1946, 35,000 military personnel were being discharged *every day*, but Washington had actually given some thought to this. To avoid the riots and recession that followed the Great War, the government instituted several phenomenally successful programs, most notably the Serviceman's Readjustment Act. Anyone, including women and minorities, who had served at least ninety days and was discharged honorably, could receive the benefits. Zero money down, low-interest mortgages got veterans out of the cities into newly constructed suburbs where 1.4 million houses were built each year after 1946. Additionally, the 52-20 provision provided $20 per week for a year in unemployment payments, though less than one-fifth of veterans ever claimed it. For a generation born in the turmoil of the 1920s, raised during the Great Depression, and having won the world war, nothing seemed to fulfill life better than an education, a job, and a house.

However, it was the educational benefits that provided far-reaching impacts for the nation. Tuition and a housing allowance were paid for veterans to finish high school, learn a trade, or, if they could be accepted, enter a university. Nearly eight million men and women took advantage of this, including two million who were now able to attend college. In 1940 there were 109,000 bachelor degrees awarded to men, and 77,000 to women nation-wide, but by end of the decade this had increased to 328,000 and 103,000, respectively. Not only did these measures prevent a repeat of the 1920s, but they created a responsible, largely stable, and educated middle class that propelled the United States into the global leadership position it still enjoys today.

During the war, America produced twice the matériel of Germany, Italy, and Japan combined, so when the conflict ended, those vast industries turned back to manufacturing civilian products. Car companies once again made cars instead of tanks, half-tracks, or jeeps; shipyards and aircraft makers expanded rapidly into peacetime, commercial markets; and, in several cases, chemical companies turned their products into fertilizers and pesticides. Aided by government subsidies and farm loans, agriculture once again became a viable business.

High-yield crops were developed and, when paired with new tractors and combine harvesters, provided enough food for America and a good part of the world. From the summer of 1945 through June 1946, American farmers produced enough surplus to export 17 million tons of wheat to Japan and Europe, thus staving off a postwar disaster and giving recovery a chance—a recovery that would hopefully prevent a descent into chaos similar to that of the 1920s, and quite possibly sow the seeds of another global conflict.

To this end George C. Marshall, former Army chief of staff and secretary of state under President Truman, proposed the European Recovery Program, later known simply as the Marshall Plan. Over 70 percent of Europe's infrastructure had been destroyed, millions of homes lost, and at least twelve million refugees were flooding into the west from the east. Industrial production was barely half of prewar levels, and food was a real problem. Worse still was the economy. Britain, though victorious, was broke. Germany and Italy essentially had no economy and both needed new currencies. There were those who, somewhat understandably, wanted the former Axis powers to suffer because, after all, their folly and aggression had killed off 3 to 4 percent of the world's population; over sixty million human lives. Franklin Roosevelt himself stated:

> Too many people here and in England hold the view that the German people as a whole are not responsible for what has taken place—that only a few Nazis are responsible. That unfortunately is not based on fact. The German people must have it driven home to them that the whole nation has been engaged in a lawless conspiracy against the decencies of modern civilization.

But there were also those who realized that the only chance for life without another world war lay in removing the causes that would instigate such a catastrophe, and in the end this view prevailed. Some $12 billion was sent overseas in the forms of loans and grants to provide food, fuel, and reconstruction; and not just to former enemies. The United Kingdom actually received the largest percentage of aid, approximately 26 percent, followed by France with 18 percent. American goods were then purchased with American money. The idea was very similar to that attempted during the 1920s, but this time it worked. Europe was spared a haphazard, plodding recovery, so within a few years industrial production was back above prewar levels, standards of living were up, and discontent largely contained.

This was not an altogether altruistic act on behalf of the United States. Decency and humanity certainly did play a part, but the overriding motives were based on national security. It was reasoned that the safest future for America lay in *not* fighting another world war, and if Europe or Asia were permitted to languish, then they would be easy prey to alternative ideologies. A familiarization with, and increasing dependency on, U.S. goods was stronger than a *military alliance, and much cheaper. America was no longer isolationist in nature, but now truly globalist and* her presence was felt in places never encountered before the war. Also, nearly 5 percent of Marshall Plan aid went to the CIA and "other" government funding abroad in an increasing effort to contain the next threat—a very real and diametrically opposed system of government: communism.

Theoretically, this was a system where ownership of everything—property, resources, land—belonged to the "community," or the people. An extreme derivative of socialism, Russian communism along the Lenin-Stalin model went far beyond economics into the political and ideological realm. In both systems, the vital means of production and distribution are controlled by the state, yet socialists reallocated these resources based on an individual's input and efforts, while a communist distributes based on needs. The fatal flaw in both methods is that *someone*, or some group, has to determine who is allocated what, and how it is done. This cuts pure communist/socialist doctrine off at the knees and negates its adherents' very argument against capitalism, which they oppose due to the perception that a small elite controls a society. In fact, a communist society is much worse off than its capitalist equivalent because there are no checks and balances against an all-powerful state. There are no incentives for working harder. Innovation and motivation are severely limited and few, if any, alternatives exist save revolution.

It was no wonder that the Western powers abhorred the Soviet Union as a political system and were equally despised in turn. Communists saw order and discipline in their system versus chaos and confusion in the West. They prided themselves on self-sacrifice for the good of all, while disdaining what was promoted as the selfish narcissism of the West. It all came down to control: whether humans have the right of self-determination or should subordinate their own interests for the common good—whatever that is determined to be by those who take charge.

The West, and especially the United States, *was* tumultuous; Communists pointed triumphantly to the Great Depression as the consequences of capitalism and the failure of a weak government to protect its people. But they were also embarrassingly wrong as billions of dollars' worth of Lend-Lease vehicles, aircraft, planes, trucks, clothing, food, and ten million pairs of boots indisputably proved.* Messy as the capitalist West was, the strength of free peoples determining their own destiny proved stronger—both militarily, technically, and creatively—than a gray, oppressive dictatorship disguised as socialism.

So it is no wonder that once the necessity of war against a shared enemy ended, the gaping ideological differences between East and West, capitalism and socialism, could no longer be ignored. The Soviet Union was immense; it possessed nearly unlimited manpower and, straddling Europe and Asia, initially had a tremendous geographical advantage. The reality of an immensely powerful United States facing an expansionist, unified Soviet Union quickly frayed the wartime alliance, and Communist antipathy toward the United States (and vice versa) rapidly resurfaced. Correct or not, these deep suspicions and fears became a rallying cry every bit as evocative as "Remember Pearl Harbor," with an anti-Communist strategy shaping America, and indeed the world, for nearly fifty years.

The USSR believed itself to be a global power and, through the spread of communism, sought to bring as much of the world under its sway as possible. Winston Churchill, eloquent and direct, phrased the situation thus: "From Stettin in the Baltic to Trieste in the Adriatic, an iron curtain has descended across the Continent."† Yet there was a wariness of the West, or at least its technology, that made Stalin hesitate. Fully 12 percent of the Red Air Force consisted of Western-supplied fighters; Spitfire Mk. Vs and Hurricanes from the British; P-40 Kittyhawks,

---

* Including millions of pounds worth of aluminum, cobalt, graphite, and heavy water for the Soviet nuclear program.

† March 5, 1946, at Westminster College.

A-20s, Airacobras, and Kingcobras from the Americans. The average Russian might believe that they alone had defeated Hitler, but Soviet leadership knew the truth, and also knew the USSR could not risk open war with the United States—at least not yet. So the slow spread of communism began and would be especially attractive to countries that had been colonized by the West, and whose people could not or would not distinguish between capitalism and colonialism. For these nations, communism seemed a viable alternative, an expression of nationalism and independence. Moscow encouraged this fallacy as much as Washington sought to contain it, and what soon became known as the Cold War began in earnest.

Military priorities and development closely follow civilian goals or diplomatic failures, and certainly the first few postwar years were no exception. The impetus to expand their sphere of influence was certainly political, but it was equally certain that this would be backed by the threat of military action, if not directly then through surrogates. On the ground the Soviets were not unduly concerned, and by the war's end the Red Army could field 500 infantry divisions, at least fifty tank brigades, and thousands of aircraft all at a time when the U.S. military was discharging thousands of men per day. Yet the Russians had learned the value of airpower from the Luftwaffe, nor would they forget that the Americans had flown over two million combat sorties from deployed locations far from home while the Red Army was still using horses and wagons.*

Recognizing both the potential of German advanced tech-

---

* In fact, the leading Allied ace was a Russian, Ivan Kozhedub, with sixty-two kills.

nology and the Soviets' own lagging position in that area, Moscow resolved to do everything possible to equalize the situation so their political agenda could progress. As the war ended, the Soviets also had teams gathering up scientists, documents, and physical technology wherever possible. In May 1945, when the Third Shock Army entered Berlin, they captured reams of DVL (German Aviation Research Establishment) information on high-speed aerodynamics, swept wings, the Focke-Wulf Ta 183, and other programs.

Never forgetting that the atomic bombs, which Russia still did not possess, had been dropped by B-29 Superfortresses capable of high-altitude, deep penetrations, Stalin ordered the creation of a simple, tough interceptor capable of stopping American bombers. Using the captured German data as a point of departure, the Mikoyan-Gurevich (MiG) Experimental Design Bureau produced the MiG-9 jet fighter, which flew in April 1946 and greatly resembled the Focke-Wulf Ta 183. It featured a straight wing, high tailplane, and nose-mounted German BMW 003 engines, which suffered from frequent failures like all wartime engines. However, the Russians learned quickly and were adept at modification, if not innovation.

Yet by 1946 the United States had two jet fighters: the P-59 Airacomet, relegated now to maintenance and pilot training; and Lockheed's P-80 Shooting Star. Both were assigned to the 412th Fighter Group and over forty Shooting Stars had been flying over the high desert since July 1945. A rash of P-80 fatalities, including Major Dick Bong's death in the late summer, brought jet safety to the forefront and threatened all future jet programs. General Henry "Hap" Arnold, chief of the U.S. Army Air Force, directed that the test and evaluation shortcuts accomplished for wartime expediency be rectified immediately.

Arnold, who had been taught to fly by the Wright brothers and was one of the original three Army pilots, was also a visionary who understood both the promise of jet technology and the value of public relations.

The issues were solved, and just in time. There were rumors that the Russians had made good use of German scientists and data and were building a swept-wing fighter of their own that would obviate America's technical lead. North American Aviation and George Welch, in the meantime, had been working on the XP-86, a next-generation, swept-wing fighter that was intended to ensure air dominance over anything built by the Soviet Union. The other tine on the strategic fork aimed at Moscow was the ability to pass the speed of sound in a manned aircraft. This capability would defeat any type of current air defense system and, when designed into a bomber, could deliver atomic weapons anywhere in the world with impunity. Such a breakthrough, whether done with a jet or a rocket, would alter the balance of power in the world, just as the threat of it would hopefully keep the peace.

═══════

Eight

The Final Stage

Y es . . . I said it is quite practical to build a plane that can
fly at a thousand miles an hour." Theodore von Kármán
stated this unequivocally when the question was posed
in 1943 by General Frank Carroll, chief of the Army's Engineering Division. The Hungarian was the head of Caltech's
rocketry program, the only one in the United States, and an
internationally recognized expert in applied mechanics. At the
height of the war, he had been invited to Wright Field for a
discussion of the practical aspects of such a plane in light of
German advances and Frank Whittle's jet engine. John Stack
and Bob Gilruth of the NACA had recorded supersonic airflow
over areas of a NACA XP-51 wing, and they verified speeds exceeding Mach 1.3 with rockets fired from Wallops Island off the
Virginia coast. Knowing it was possible for a craft to exceed the
Mach, and now well aware that such flying represented no in-
surmountable barrier to future development, American technology came to a crossroads.

A major issue was that of data. Pistons and props were of no use in transonic evaluation because conventional aircraft only got to this point if they had exceeded their design limits, and there was no hard information for this flight region. Test flight data was largely anecdotal and, even when the pilot survived, it was uninstrumented and therefore not verifiable. North American Aviation's Mustang was the exception, as recording equipment had been placed in the wing's gun bay.

In fact, it was with this very XP-51 that the shock wave phenomena was first observed. When an object (for our purposes an aircraft) moves, it disturbs the air and this disturbance spreads, or propagates, ahead of the aircraft. The "wave," or pressure field, caused by this pushes the air ahead out of the way and will continue to do so as long as the plane remains in the low-speed, subsonic region. Also, as an aircraft exceeds approximately 0.7 Mach it enters the transonic region and is traveling nearly as fast as the wave itself. There is no longer a pressure field out front to move the air aside, so all the pressure and velocity changes occur suddenly and unpredictably. Air molecules now cannot be moved until the aircraft itself moves them, and the result is called a *compression*, or *shock wave*, which forms at the aircraft's leading edge.

Such a wave creates an area of high pressure directly behind it, and with this greater pressure comes a drastic increase in drag. The effects of supersonic wave drag generate the most significant differences between subsonic aerodynamics and those of an aircraft flying faster than sound. Current wind tunnels were next to useless because as the model became transonic, the generated shock waves bounced off the walls and back onto the model, which negated any meaningful measurements. This problem would eventually be solved by Ray Wright of the

NACA, who added slots, like horizontal vents, into the tunnel, thereby allowing the shock wave to disperse. Meanwhile, the only way to obtain accurate data was with an instrumented test aircraft flown by a test pilot, and, significantly, NAA had the head start.

John Stack wanted to build a turbojet to study sustained transonic flight and solve those unknowns. He, and others like him, reasoned that while a rocket could punch through the Mach, as they had done from Wallops Island, there was very little practical use for such aircraft. Rocketry had a future for defense systems and perhaps for blasting beyond Earth's atmosphere one day, but, as the Germans had discovered, military applications were severely limited. Frank Carroll and Ezra Kotcher, one of his trusted engineers, wanted the rocket precisely because it could muscle its way through the transonic region and into supersonic flight. Kotcher had been a proponent of this after attending a lecture given by Hermann Zornig, a ballistics expert, who authored an influential paper on the subject.[*]

The Army, who ultimately held the purse strings for funding, agreed with Carroll. During the war there was no time for nonmilitary research and resources were too valuable to expend this way. Bell's XP-59 was receiving lukewarm reviews, at best, so the jet was not necessarily the answer and was likely the safest option. A rocket would also provide a higher thrust-to-weight ratio as it was lighter than a jet, and this meant the chance of punching through Mach 1 was probably better.

---

[*] Zornig is frequently overlooked, but he compiled a great deal of supersonic data while chief of the Army's Ammunition Division. He also founded the first Scientific Advisory Committee to the Ballistic Research Laboratory.

They went with the rocket.

And, somewhat surprisingly, with Bell Aircraft, though with hindsight the decision seemed logical. General Hap Arnold was pleased with what the Airacomet represented, and he personally liked Larry Bell, who had the only real experience manufacturing advanced technology with the XP-59. Also, the other companies and their subcontractors were stretched thin enough building workable combat aircraft while Bell was not. So while the Germans tried to blitz through the Ardennes and the Battle of the Bulge reached its climax, the Air Corps, Bell Aircraft, and the NACA formally agreed to the draft specification for a manned, single-seat research aircraft capable of stable flight up to 0.8 Mach. It would be fully instrumented and able to withstand forces eight times greater than gravity. In March 1945 while Chuck Yeager was on his honeymoon and Ken Chilstrom, now operations officer for the Fighter Test Section at Wright, was testing Bell's Airacomet, official contract W33-038-ac-9183 was signed for the Experimental (X) Supersonic (S) aircraft number One: the XS-1.*

Captains Fred D. Orazio and G. W. Bailey of Wright Field's Design Branch drew up the initial plans under Ezra Kotcher's close supervision. Bell's ballistics research on the .50-caliber bullet had shown stability in both the transonic and supersonic regimes (otherwise, bullets would be hopelessly inaccurate) so this provided the basic fuselage design. Its wings were straight since the German swept data had not yet been captured and Larry Greene, Bell's chief designer, placed the horizontal tail surface on the craft's vertical tail, not the empennage. He also

---

* The "S" was quickly dropped and it became universally known as the X-1.

added a method by which the angle of the horizontal tail could be manually adjusted from the cockpit by the pilot—very similar to the system employed by the Me 262.

One of the issues worked out in 1945 was whether the X-1 would perform a conventional takeoff or be air-dropped from a mother ship. Looking ahead toward future development options as an interceptor, Bell's initial position was to make the plane capable of a normal takeoff, but the fuel consumption numbers did not add up. The extra weight for larger tanks and more fuel would have precluded the aircraft from getting to its test altitude. Additionally, a turbopump was required to move the rocket propellant from its storage tanks to the combustion chambers, and this pump never quite functioned correctly. The only immediate solution was using nitrogen to force the propellant transfer, and once these additional tanks were installed the loss in fuel capacity, and extra weight, ruled out any other option than an airdrop.

Bell delivered its first full-scale mock-up in October and there were no issues, so on December 27, 1945, X-1 #46-062 emerged from the company facility in Wheatfield, New York. Flown by a B-29 bomber to Pinecastle Field near Orlando on January 19, 1946, the rocket plane completed preliminary low-speed testing and was ready to fly six days later. Jack Woolams, a former Army Air Corps pilot and now Bell's chief test pilot, was dropped from a B-29 at 27,000 feet over central Florida on January 25, 1946. Its short, four-minute glide flight was uneventful and Woolams later wrote that "the airplane felt as solid as a rock, experiencing absolutely no vibration or noise. Longitudinal stability is quite positive . . . and lateral stability is about neutral." During the next five weeks, Woolams completed ten glide flights with the X-1, and small modifications were made before the aircraft moved west to Muroc for the final phase of the program: intentional, manned flight beyond the speed of sound.

Nineteen forty-six, then, dawned with hope and promise. "It was a banner year," Ken Chilstrom remembers fondly. "It was actually the best year for me as a pilot!" Two days after the X-1 glide flight, Colonel Bill Councill lifted off from Daughtery Field in Long Beach, California, turned east, and streaked 2,457 miles across the United States in a Lockheed P-80A-1 Shooting Star. Maintaining an average speed of 584 miles per hour, he made it nonstop to New York's La Guardia Field in four hours, thirteen minutes, and twenty-six seconds, thus setting a new Fédération Aéronautique Internationale record.* Councill's feat was significant for several reasons; first, it demonstrated the jet's capability to safely cross great distances at high speed, and this was not lost on military planners from either side of the Atlantic. Second, the flight fired the imagination of a war-weary public and gave a very clear, exciting look into the future of aviation.

With the most recent war won, the U.S. government and military leadership now seemed determined to never be caught unawares as they had been in 1941 so, to that end, development and testing of new aircraft would include as many combat veterans as possible. Men who had seen what works and what did

---

* Two other P-80s piloted by Captains John Babel and Martin Smith flew with him, but did not fly nonstop as they were forced to refuel in Topeka, Kansas. In 1927, Charles Lindbergh made roughly the same trip in fourteen hours and twenty-five minutes.

not, who knew what was important beyond the drawing boards and the boardrooms. Consequently, Fighter Test at Wright Field was now home to many legendary Air Corps pilots from the war. Men like Richard Bong, Don Gentile, "Gabby" Gabreski, and Gus Lundquist. Also pilots like Ken Chilstrom and Glen Edwards who had fought the close-air-support side of the war, and younger pilots like Bob Hoover or Chuck Yeager, who were eager to fly anything they could get their hands on.

Through the spring of 1946, Bell worked out the X-1's bugs and Chilstrom's boss, Colonel Al Boyd, picked him for a special mission that, quite pleasingly, closed a loop in aviation history. On the surface, it seemed an annoyance but Ken, who admired Al Boyd like no other man, would have done anything asked by the chief of Flight Test. On June 22, 1946, just three years after he flew a piston-engined fighter into combat over Pantelleria, Ken piloted a Shooting Star jet fighter up to Schenectady, New York, and picked up a letter from General Electric's corporate office. GE, of course, had been granted the initial U.S. license to manufacture Frank Whittle's W.1 engine during the war and continued its progress with the I-16 and I-40 jets. Ken's P-80 was powered by a J33-GE-11 jet engine, and the letter was one of thanks that the Air Corps wanted hand delivered: to none other than Orville Wright, and at the very field that bore his name.

"I landed at Wright and taxied slowly up to the base of the tower," Chilstrom reflected nostalgically. "There was a black sedan there, all alone. I shut the jet down, climbed down, and pulled the letter out of my chest pocket. It was a little sweaty." He chuckled. "Anyway, I walked over to the car, and the back window rolled down and there he was—Orville Wright. I thanked him on behalf of the Air Corps, and hoped maybe he'd want to look at the jet, but he never said a word. Never smiled or

said anything back. The window went up and he drove away."
Nevertheless, an aviation circle had been completed. The man
who struggled into the air off a rail on a sand dune got to watch
a sleek, all-metal jet fighter land on the very field where he and
his brother had flown open-cockpit, fabric-covered biplanes.

Ken could well afford to dismiss Orville Wright's rudeness;
he had other things to do, because he was now the acting head
of the Fighter Test section. There were foreign aircraft to fly,
developmental jets, and of course word had filtered down about
two new programs: Bell's X-1 and the XP-86 from North American Aviation. That summer Glen Edwards was already heading
to Princeton for a master's degree in aeronautical engineering,
and Gus Lundquist was finishing up at Duke. Chilstrom got
the official word to "get three P-80s ready for the National Air
Races at Cleveland. It was big deal . . . this was a race Jimmy
Doolittle had flown in."

Indeed it was.

Originating in 1920 with Joseph Pulitzer, the race had gone
through many iterations, locations, and names in the past
twenty-five years. It had thrilled spectators during the Roaring Twenties and inspired the public during the lean years of
the Great Depression. With war looming, the "300 miles of
the world's toughest flying" gave way to the truly tough flying
of combat, and the last race was flown in 1939. Now, seven
years later, some ninety pilots came together again with turbo-
charged piston-engined surplus fighters from the war and, for
the first time ever, the jet.

Following Bill Councill's flight, and the exotic, exciting developments in the skies over California and Ohio, a special jet
category was added to the race. Only the military had the new
aircraft, so it would be an all-military lineup. Three P-80s from

Wright Field Fighter Test would compete against three operational Shooting Stars from the 1st Fighter Group at March
Field. It was an ideal venue for the Air Corps to showcase its
newest weapon and, perhaps more important, to garner official
support (and funding) for additional advanced programs.

Not only did Ken Chilstrom get to choose the specific jets,
but he also picked the pilots. "Well, myself of course." He
smiled. "I had more jet time than anyone else right then and in
those days if a dangerous mission came down, then the guy in
charge was supposed to take it." Each engine was handpicked
then carefully primed to get every pound of thrust possible.
The guns were removed, and all exterior ports sealed to reduce
drag. Also, Wright engineers analyzed the parallelogram race
circuit and worked out the optimum airspeeds and g's for the
turns. Chilstrom decided to fly P-80 #44-85123, the same jet
Bill Councill used to set the nonstop, coast-to-coast speed record back in January.

But Major Gus Lundquist returned from Duke that summer
and, as the military is fond of saying, rank has its privileges.
Gus was now also Ken's boss, so no one thought anything of it
when he decided to lead the Wright team, and take the #123 jet.
"That's just how it goes. I smiled and nodded," Ken recalled.
"Gus was my friend, and a helluva good pilot . . . the finest test
pilot I ever knew." Chilstrom would fly #044, and Captain J. E.
Sullivan would be the third pilot with P-80 #247.

Out in California, the three 1st Fighter Group Shooting Stars
would be led by Major Robin Olds, West Point grad and World
War II ace. Arguably the Air Corps's finest active-duty pilot,
Olds despised bureaucracy, politics, and rear-echelon desk
jockeys; he was a warrior through and through. "Everyone who
mattered liked Robin Olds," Chilstrom stated. "He was a fierce

competitor and a fighter pilot's fighter pilot. Olds was a combat squadron commander at age twenty-two and had twelve confirmed kills in the ETO. His men would follow him anywhere, women adored him, and he married a movie star."*

The operational pilots had all their engines tuned to run at better than 100 percent, a dangerous option as the turbines would overheat, but they were every bit as determined to win as the test group. In the midst of all this, the X-1 program suffered a tremendous blow. Bell's Jack Woolams was planning to race in the piston-engine category in Cleveland, and the company modified a P-39 Airacobra for the event. On August 30, just days before the race, the plane came apart over Lake Ontario and Jack was killed.

However, all test flight programs are more than "one deep," so Chalmers "Slick" Goodlin became Bell's new chief test pilot and head of the X-1 flight tests. Slick had been flying since he was a teenager and in 1941, at age eighteen, joined the Royal Canadian Air Force since America was not yet in the war. By late 1942 he was in Europe flying Spitfires, but was then asked by the U.S. Navy to transfer his commission, which he did. Joining their test program, he left the Navy in late 1943 to join Bell.

So over the 1946 Labor Day weekend pilots and planes converged at Cleveland. Goodlin was involved in the race as the co-owner of a P-63 Kingcobra, though he did not fly it himself. George Welch was also there in a P-51D, but was forced out on his third ten-mile lap due to engine trouble. Ken had gremlins

---

* Olds had at least four jet kills as the 8th Tactical Fighter Wing commander during Vietnam. He was married to actress Ella Raines, who had two stars on the Hollywood Walk of Fame.

too. The Shooting Star had a compact, well-designed cockpit, and it was not unlike the A-36 in those respects—but the jet was sleeker, so the primary panel had fewer gauges, and they were conveniently grouped in threes across his field of vision.

"I'm coming around on the last leg of the race, at 515 mph, in a 60-degree bank right in front of the grandstand," he recalled. "What a perfect place for something to go wrong . . . and it did."

Suddenly the stick refused to move. In a heartbeat his entire body tightened, and fear shot up through his gut to squeeze his chest. The jet was still throbbing and there was no lurch or bump, and the earth was still zipping past. Instantly, Ken's eyes dropped to the big tachometer on the panel's right edge, then down to the engine temperature . . . slapping the throttle back to IDLE he used both hands to yank the stick left and backward . . . he had to climb! Just yards off the ground at 750 feet per second gave him no time at all to live. He felt the g's push him into the seat as the P-80s nose lifted and the ground fell away. Still in a right bank, the P-80 zoomed up and away from the crowd and Ken's eyes danced around the cockpit; no fire light . . . and squinting at a small gauge on the bottom of the pedestal he saw normal hydraulic pressure. Jets flashed past beneath him but there was sky above and he was still soaring away from the people and the unforgiving ground.

No pilot ever won a game of chicken with the earth. Sunlight filled the cockpit and he could smell the fighter . . . hot metal and warm leather from the harness. Sweat rolled over his nose, but he started breathing again and tugged the throttle back a few inches while he figured out what in the hell just happened. Due to the g-forces exerted because of its increased speed, the jet's ailerons had to be hydraulically boosted, and the pump that did this failed on Chilstrom's jet. Essentially freezing the stick, roll control was instantly impossible, though the elevators still worked.

"Kinda eye opening at that speed a few hundred feet off the ground," Ken remembered. "So I chopped the power, pulled back, and zoomed up out of the race. I could see Gus and Robin wingtip to wingtip around the last few turns. Once I got below 200 mph, the ailerons worked manually so I could land . . . but I was out of the race." Lundquist won it for the Wright Field test pilots, and Robin Olds placed second. Another 1st Group pilot, Captain A. M. Fell, came in third and Captain Sullivan of Wright was fourth. One of the operational pilots, a lieutenant colonel named Petit, was disqualified for cutting a pylon, and Ken was out for his boost pump problem. Incidentally, Robin Olds was so short of fuel he could not taxi back in, and the jet flamed out on the tarmac in front of the grandstand.

Nevertheless, it was a great day to eat popcorn, drink an icy Coca-Cola in the sun, and thrill to the sight of fascinating aircraft racing across the sky. There was no bad news from far-off places, and no War Department telegrams beginning with "regrets to inform you . . ." delivered by Western Union couriers. City lights glowed at night, there was gasoline for everyone, and pilots were dashing heroes inspiring children and reminding grown-ups how good it was to be American. Not that they needed such reinforcement a year after the war ended, but, as Ken Chilstrom summed up that summer, "It felt good to feel good again!"

But the world is never completely docile and despite, or maybe because of, the war's end, 1946 was no exception. Earlier in the year the United Nations held its inaugural assembly in London and within weeks there were issues. The Soviet Union had occupied large swaths of northern Iran, ostensibly to protect their southern flank from the Germans. In reality, the move had everything to do with securing vast oil deposits, threatening British interests in Abadan, and expanding the Communists' sphere of influence. Washington supported Iran against the encroachment and used the UN to force the Soviets back.

Moscow was, perhaps understandably, furious. The bald hypocrisy of the issue, which would reoccur frequently over the years, threatened UN credibility from the onset. After all, the Russians reminded everyone, Britain still retained most of its empire while France, which had done nothing to ensure the final victory, similarly retained Indochina, parts of North Africa, and was currently occupying Lebanon and Syria. The Soviet ambassador stormed out on March 26, and the world was fast becoming polarized again. This time it was ideological aggression; Communist socialism against everything else. Actually it was the same struggle: groups of men on both sides who sought to control millions of destinies through a dangerous combination of patriotism, ambition, conviction, and fear. Moscow was adamant and so was Washington; in fact, the USS *Missouri* was ordered into the Black Sea by way of the Turkish Straits, to prove the point.

In the United States this new fear manifested itself in the form of another "Red Scare." Anything or anybody remotely suspicious could be, and often was, branded a Communist. Spies were everywhere, the people on both sides were told; and in America, suspicion became so bad that the Federal Bureau of Investigation hired an additional 7,000 agents. For perceptive military officers and government civilians the foreign flash points were also becoming obvious; namely, Korea. The problem, like many others, was left over from the war. All territory controlled by the Japanese was parceled out between the victors, and the division of the Korean peninsula became a prime example of the often well-meaning, but fundamentally ill-informed and arrogant diplomacy characterizing so many Cold War issues. In this case, inexplicably, two relatively junior officers, Charles Hartwell Bonesteel from the Pentagon's Strategic Policy Committee and Dean Rusk from the Department of State, were allowed to arbitrarily divide Korea.

With no knowledge of the country, and without consulting anyone who knew better, they used a *National Geographic* map to bifurcate the country along the 38th Parallel; north of the line would be a Soviet Communist zone, and the south would initially be controlled by the Americans. Conflict, given the belligerency and suspicions on both sides, was inevitable. This led to a series of efforts that shaped the policy, science, culture, and geopolitics of the Cold War. It also started an unprecedented arms race that began with the atomic bomb and was accelerated by the quest for supersonic flight. In light of the world situation, pressure to be first took on new significance. The U.S. government therefore needed a place far from prying eyes, and easy to secure, so Bell moved its operation to Muroc for completion of the supersonic flight program.

On December 9, 1946, X-1 #46-063, was dropped from a B-29 over the Mojave Desert, and as the little rocket plane fell clear of the big bomber Slick Goodlin blinked against the sunlight flooding through the top of the cockpit. Unlike a normal aircraft, there was nothing in front of him but a large panel, and forward visibility was extremely limited. Gripping the big H-shaped yoke, he ran his eyes over the switches and gauges as the aircraft glided into open air away from the bomber. There was also no throttle; thrust was controlled by flipping up each of the four rocket motor toggle switches across the middle of his instrument panel and this was the point of today's test.

Jack Woolams, the previous test pilot, had only glided the X-1 but today Chal Goodlin took a deep breath and, as the San Gabriel Mountains slipped past in the corners of his eyes, he tightened his hands on the yoke and fired the first rocket. As the first chamber of the Reaction Motors XLR11 motor lit off the X-1 began its first powered flight. Using a mix of ethyl alcohol, water, and liquid oxygen (LOX), each motor was capable of producing 1,475 pounds of thrust. When all four were activated, that was 5,900 pounds for roughly four minutes, a relatively long burn made possible by a regenerative cooling method that allowed compressed gas to expand then remove the motor's heat as it cooled.

By spring 1947, several other significant events had occurred. The original X-1, #46-062, had been modified for powered flight and was being prepared for transport out to Muroc. Bell and Chalmers Goodlin needed to conduct at least twenty additional flights by midsummer in order to fulfill the company's contract. Then on June 19, Colonel Al Boyd, now chief of the new Air Materiel Command's Flight Test Division at Wright Field, set an official world speed record by reaching 623.738 mph in a racing version of the Shooting Star.* Though Luftwaffe pilots flying the Me 163 and Me 262 had gone faster during the war, there was no officially accepted proof of this, and the previous world record of 615.78 miles per hour had been set in September 1946, by a Gloster Meteor F.4 flown by Group Captain Edward "Teddy" Donaldson. The modified J-33-A-21

---

* XP-80R # 44-85200.

engine used water-methanol injection, which produced over 5,000 pounds of thrust, and as with the air race jets, all ports were sealed and the guns removed. In a precursor of things to come, the wings were also shorter and sharper than on a production model.

For certain, Al Boyd was no deskbound, paper-pushing officer. Born in Rankin, Tennessee, in 1906, Boyd finished college and was commissioned in 1929 just months before the stock market crashed. Serving as an instructor pilot, he scratched out a career during the grim years of the Great Depression at bare bases like Brooks and Kelly Fields. Managing to escape Texas, he was assigned to Hawaii in 1939 and was the Air Corps chief engineering officer when the Japanese attacked. By 1945, Boyd had been named the acting chief of the Flight Test Division for the Air Corps Air Technical Service Command (ATSC) at Wright Field. When the ATSC became the Air Materiel Command in 1946, Boyd was subsequently confirmed as the chief of Flight Test. A no-nonsense sort of officer, Al Boyd had a dry, barely discernible sense of humor and, as a first-rate aviator himself, was greatly admired by his pilots.

Through the fielding of operational P-80 squadrons, world speed records, and the National Air Race, the jet was now accepted as the future of combat aviation. Boyd's Flight Test Division had expanded tremendously and reorganized by 1947, with Ken Chilstrom taking over from Gabby Gabreski as the chief of the Fighter Operations Section. They had completed most of the exploitation of Hal Watson's captured aircraft and verified vast amounts of German aerodynamic advances. Both Adolf Busemann and Alexander Lippisch had been brought to the United States, and Chilstrom and Lundquist worked extensively with the latter at Wright Field. Busemann was quite happy to be in America, and his assistance to the NACA with their swept-wing research was invaluable. Lippisch was a different story. "Dr. Lippisch's attitude was belligerent," wrote Nate Rosengarten, a Wright Field project engineer. "He was reluctant to help us, and I believe he did not want us to fly the Me 163." Lippisch also rather implausibly denied that he could not translate the Komet's takeoff sequence for the test pilots.

The swept-wing data was invaluable, especially after North American realized it had to change the design of its new XP-86 prototype. Originally designed with a straight wing and NAA signature bubble canopy, with a top speed of 582 miles per hour it did not offer a significant performance advance over the P-80, or the Navy's FJ-1 Fury. The GE J-35 jet engine put out 4,000 pounds of thrust, but the critical Mach number still hovered at 0.8 Mach. There was some internal North American resistance to the change and, incredibly, to using any German engineering, but this was overcome in the name of reality. The NACA provided extensive wind tunnel test results, and the Flight Test at Wright gave NAA its Me 262 exploitation results, including an actual wing.

This was vital because one major drawback of the swept design is its poor low-airspeed performance. This was critical because slow-speed, high-angle-of-attack maneuvering is essential in a twisting, turning dogfight as the airspeed bleeds off, and absolutely necessary during takeoff or landing. The German solution was to add leading-edge slats, which they had successfully developed for the Bf 109, which slid out at low speeds to increase the wing's camber; in effect, to temporarily increase the wing area. A simple idea, the slats were held in place at high speeds by ram air, and as the plane slowed down this force obviously lessened so the slats moved out. There was an interlock in place, and when the gear came down, the slats automatically extended. The pilot could also manually lock them so there would be no asymmetric extension during a dogfight.

North American's first seven XP-86s basically copied the German design, though later this was improved significantly. By late 1946 it was also decided to sweep the vertical and horizontal tails, which resulted in a net improvement from 0.8 to an astounding 0.9 critical Mach number, and an increase in top speed by at least 70 miles per hour. This, plus NAA's laminar flow wing, made the XP-86 extremely "clean" and very, very fast. It was rumored that the Mikoyan-Gurevich design bureau had developed a similarly lethal swept-wing fighter of their own to replace the MiG-9, so it was vital for the United States to counter this.

The global situation, which always influences military development, had indeed become more volatile. After the *Missouri* made its startling appearance in the Black Sea, which the Soviet Union had long regarded as a Russian lake, Moscow pressured Turkey to restrict access through the straits—for security, it was claimed, and regional stability. Ankara refused. The 1936 Montreux Convention ceded control of the Dardanelles and Bosphorus, which together compose the Turkish Straits, to Turkey. Free transit of all civilian vessels during peacetime was guaranteed, but the passage of warships remained at Ankara's discretion. The United States obviously supported this decision while the USSR vehemently opposed it. Of course, any support of Turkey irritated the Greeks, Turkey's longtime nemesis, and the Soviets used the incident to make overtures in that direction. This incident and others, including Moscow's refusal to endorse the Baruch Plan for international (UN) control over atomic weapons, prompted the American president to articulate what became known as the "Truman Doctrine." In part, this extraordinary policy that molded world affairs, and still does, read:

> No government is perfect. One of the chief virtues of a democracy, however, is that its defects are always visible and under democratic processes can be pointed out and corrected.
>
> One of the primary objectives of the foreign policy of the United States is the creation of conditions in which we and other nations will be able to work out a way of life free from coercion. This was a fundamental issue in the war with Germany and Japan. Our victory was won over countries which sought to impose their will, and their way of life, upon other nations.
>
> The peoples of a number of countries of the world have recently had totalitarian regimes forced upon them against their will. The Government of the United States has made frequent protests against coercion and intimidation in violation of the Yalta agreement in Poland, Rumania, and Bulgaria. I must also state that in a number of other countries there have been similar developments.
>
> I believe that it must be the policy of the United States to support free peoples who are resisting attempted subjugation by armed minorities or by outside pressures. I believe that we must assist free peoples to work out their own destinies in their own way.
>
> The seeds of totalitarian regimes are nurtured by misery and want. They spread and grow in the evil soil of poverty and strife. They reach their full growth when the hope of a people for a better life has died. We must keep that hope alive.

*The free peoples of the world look to us for support in main-
taining their freedoms. If we falter in our leadership, we may
endanger the peace of the world. And we shall surely endanger
the welfare of this nation.*

In other words, the United States will intervene anywhere in
the world it chooses in order to defend those who cannot defend
themselves, and in the name of national security. An admirable sentiment and generally well-meaning though it could be
construed as, and indeed was used as, justification to advance
Washington's interests. The Soviet Union, and later Red China,
North Korea, and North Vietnam, along with the balance of
the Communist world, would interpret this as a threat to their
own interests and security. Applied incorrectly, this policy was
certain to involve the United States in actions it would not have
otherwise taken. It did, in fact, codify the differences defining
the Cold War, though most did not view it as such at the time.

Facing this new, global reality would call for a different sort
of military than the one that fought the Second World War. As
President Truman explained to Congress, "whether we like it or
not, we must all recognize that the victory which we have won
has placed upon the American people the continuing burden
of responsibility for world leadership." Gone were the isolationist days of the 1930s, and the immense, multi-theater military
capability of the war years was insupportable by a democratic
constituency and financially unsustainable. Through technology and maintaining the nuclear edge, it was thought that
the Truman Doctrine was feasible. Airpower had indisputably
proven its value, both tactically and strategically, so Washington envisioned this as the way to project American power and
influence while maintaining a reasonably compact military.

"We must never fight another war the way we fought the last two," Truman informed his staff. "I have the feeling that if the Army and the Navy had fought our enemies as hard as they fought each other, the war would have ended much earlier." There was a great deal of truth in this for, despite the often desperate situations, lives at stake, and dire consequences for failure, there remained fierce competition between the Army and Navy. Battlefield success amplified the respective merits of each service branch and translated into personal accolades for their leaders. This is certainly not to say that the fighting admirals and generals were motivated by glory; the best of them were not. But those in Washington were looking ahead to the postwar era and that magic pot of gold under the political rainbow: funding. Money was everything, especially research and development projects that became new weapons, and subsequently beget new missions.

The war had shown the politicians the effects of a military principle known as "unity of command." Combat is confusing enough without having to figure out who is in charge, and this had to be extended back to Capitol Hill; otherwise, the services would continue operating independently of each other and with their own interests at heart, not necessarily national interests. MacArthur's drive across the Pacific and his postwar actions are a superb example. It was also recognized that such unity needed to encompass all the tools for making war or national defense; this included intelligence gathering and procurement.

There was opposition. The Navy had always feared the Army's political power and was afraid a reorganization/unification of the military would cost it the Marine Corps and Naval Aviation. There were also those in Congress who believed this move would "Prussianize" the United States by creating an enormously powerful military elite. To allay these fears, Navy Secretary James Forrestal commissioned Ferdinand Eberstadt, an old friend, to research the issues. Eberstadt produced a remarkable report that, along with the man himself, has received very little historical credit.

Essentially, it advocated the creation of a National Security Council (NSC) rather than outright military unification. There were too many differences in military roles, he said, too much valuable tradition, and too large a diversity in requirements for a single military. Instead, Eberstadt proposed the NSC would formulate the policies based on foreign threats, domestic defense, and future requirements. There would be military members on the council, but also the secretary of state and the president. At no time, most agreed, would the U.S. fundamental principle of civilian control over the military be abrogated. General of the Army George Marshall was adamant about this point, and the danger of permitting military dominance of foreign affairs— now more than ever.

In the end, a compromise was reached and the National Security Act was signed on July 26, 1947. Besides the NSC, this act created the Central Intelligence Agency for the coordination (theoretically) of all intelligence functions necessary for the nation's security, with a mandate that one of its two top individuals would always be a civilian. The armed forces would be reorganized, but with a civilian secretary overseeing each branch and serving as a liaison to Congress. The Navy would retain the Marine Corps and its Naval Aviation branch due to the unique challenges it faced; the war in the Pacific bore out the good sense in this. The Army would remain the army except, in an evolutionary move signaling the shift in priorities, its Air Corps would become a separate service: the United States Air Force.

Though he was a career Army officer, Ken Chilstrom knew this was the right path for a service that had won its spurs in combat. "We never did think much like the infantry or tank fellas," he remembers. "They were mostly very traditional and skeptical of new technology . . . unless it was for bullets or guns. The separation had to happen, especially given what was going on with the Russians. This is why we knew the jet was the future . . . and the new Air Force was going to show that to everyone."

Yet the new service faced an uphill climb in some areas, as the Army and Navy jealously protected their budgets and regarded the upstart U.S. Air Force as a threat. The inaugural secretary of the Air Force, Stuart Symington, needed something to demonstrate the value of the new service. Something no one else had done or could do, a capability that would simultaneously alter the international balance of power in favor of the United States, fire the public's imagination, and open funding streams from Congress. In fact, there was one program about to be transferred to the new Air Force that he knew fit all the requirements: the X-1 and the achievement of supersonic flight.

By May, Slick Goodlin had competed the requisite powered flights, verifying the contractual requirements that the aircraft withstand eight g's and was stable to 0.8 Mach. Contrary to Hollywood's story, he never refused to fly because he had not been paid $150,000. In a 1989 interview with *Air & Space Magazine*, Goodlin was understandably bitter: "That account is false. I had a handshake deal with Bob Stanley of Bell that I would make the first supersonic flight before we turned the plane over to the Air Force. He agreed I'd get $150,000 for the supersonic flights." Bonuses paid to test pilots were not uncommon, but the provision for this amount in Bell's proposed contract to continue supersonic flight testing came at a time when budget cutbacks were very real.

The overall timing fit well with Symington's ambitions so in June 1947, the X-1 was officially turned over to the Army Air Corps and the AMC Flight Test Division, specifically, to Colonel Al Boyd and Lieutenant Colonel Fred Ascani, his deputy. Born Alfredo John Ascani in Beloit, Wisconsin, Ascani was one of those boys whose life was changed forever by Charles Lindbergh. Seeing the *Spirit of St. Louis* fly over on its way to New York before leaving for Paris inspired him as nothing else had. Graduating from West Point in 1941, Ascani's class standing permitted him to choose his branch, and he chose the Air Corps without hesitation.

Wanting fighters, he volunteered as an instructor pilot, hoping this would improve his chances, but in 1943 there was a huge demand for bomber pilots due to the horrible casualties in Europe. Ascani was sent to B-17s and then on to command the 816th Bombardment Squadron out of Foggia, Italy. Returning home after fifty-two combat missions, he was assigned to Wright Field, and the Bomber Test Section where he would shortly become the Flight Test Division deputy. When the X-1 program came along, Boyd could not decide which pilot to choose so he asked his deputy. There was a lot riding on this, and much that could go wrong, not only aerodynamically but politically. According to Ascani, when Boyd asked for a pilot to take the program he only considered one pilot. "My own choice would've been the easiest for him to make," Ascani later wrote. "Major Ken Chilstrom, head of the Fighter Test Section."

Ken, or "K.O." as he was often called, was extremely smart, very smooth, and his combat experience made him exceptionally calm under pressure. The problem was that Boyd wanted Chilstrom to head up the Phase II tests of the XP-86 program, which, in most everyone's opinion, would be the finest jet aircraft yet produced. Without a viable, swept-wing fighter to oppose the new MiGs, it would not matter how many rocket ships America built. Chilstrom was essential for this and was the perfect pilot to see that vital program through. As engineers and pilots, both Boyd and Ascani knew Mach 1 could be exceeded by any reasonably competent pilot, whereas the XP-86 program represented the immediate future of the Air Force. "If you want it," Al Boyd said to Chilstrom referring to the X-1, "then it's yours." They decided to leave the choice to Ken.

"I didn't want it," he recalled, shaking his head adamantly. "I didn't think much of Bell's aircraft or of them as a company. Their war record was lousy and their planes, well . . ." He spread his hands and shrugged. "I was looking to the future and that was the jet and the XP-86, not the rocket. The X-1 program . . . who knows? Others used the phrase 'sound barrier' but that was a misnomer. It did not exist and we all knew it."

So Ascani asked Chilstrom for a list of pilots from the Fighter Section who could do the X-1, and Ken gave him seven names, among them Bob Hoover and Chuck Yeager. "They were junior guys," he says. "All of them were exceptional pilots, but not really heavily involved with other projects so they were, well, expendable."

Ascani and Boyd both liked Yeager. They agreed he was a top-notch, instinctive pilot, and both he and Hoover had more balls than sense. "Education was not a factor," Ascani recalled, "or else Yeager would have been quickly eliminated. Chuck was very unpolished. He barely spoke English. I'm not referring to his West Virginia drawl; I mean grammar and syntax. He could barely construct a recognizable sentence."

Boyd agreed and still could not decide. If the new Air Force needed the X-1 to prove supersonic flight was possible, then whoever flew it needed to adequately represent the service. According to Fred Ascani, "Boyd fretted about the Air Corps's image if its hero didn't know a verb from a noun. He decided that before Chuck would meet the public, I would give him English lessons." In the end it did not matter if the X-1's military pilot really understood the aerodynamics around him. Bell and Chal Goodlin had already fleshed out many of the preliminary details, so as long as the Army test pilot could control the aircraft through the transonic region and past the Mach, then the flight had a good chance of success. An instinctive stick-and-rudder pilot like Yeager could probably do just that.

There was more to it than image, though, so if Yeager was chosen, then both Boyd and Ascani agreed that someone would need to be along to coach him through the aerodynamic and engineering aspects of the aircraft. In the end, Boyd knew Chilstrom had to work the military Phase II tests on the XP-86; Yeager was too inexperienced for that, so Chuck, and Bob Hoover as a backup, would be assigned to the X-1 with Jack Ridley along to keep the pair out of trouble.

"Well, . . . Hoover and I were definitely not flight test engineers!" Yeager admitted later. "We could fly airplanes . . . but Jack Ridley . . . was a brain! Jack Ridley knew everything there was to know about aerodynamics and he was practical. And, besides, he was a good pilot. He spoke our language. Bob was a Tennessean and I was a West Virginian and, being an Okie, Jack spoke real good language for us."

Ridley gained a degree in mechanical engineering from the University of Oklahoma in 1939 and earned his Air Corps wings by 1942. Because of his academic background he got trapped into accepting aircraft for the military from various civilian contractors. Unable to get to a combat unit, Jack attended the Army Air Force School of Engineering at Wright Field then, in 1945, went to Caltech for a master's in aeronautical engineering. Returning to Wright Field, he was exactly what Al Boyd wanted: an educated project engineer who was also a pilot. It was Ridley's job to translate the aerodynamics to Hoover and Yeager and offer any improvements for the X-1 design.

The day after the transfer there was a meeting, hosted by the military, at Wright Field, to discuss any overlap between the NACA and military programs and sketch out the fastest, safest path to exceed Mach 1. It was decided that the first aircraft, #44-062, would be flown at Muroc because the thin wing raised the critical Mach number, and this offered the best chance for transonic-to-supersonic stability. The second X-1, #44-063, with its thicker wing, would be used by the NACA for methodically gathering transonic flight data. Telemetry for both aircraft would be provided by the NACA, and the agency would have access to all collected data. The additional 500-pound instrumentation package was truly impressive for 1947; among other things this included a five-channel transmitter for all airspeed, altitude, and acceleration readings; 400 distribution orifices to take exact measurements of transonic and supersonic pressures; and a camera to film the main cockpit panel.

The level of cooperation between a contractor, a government research agency, and the military was unprecedented and very promising. Walt Williams, the NACA site engineer in charge, served as liaison. But Colonel Al Boyd left no doubt that this was now a military program and, in light of current world events, there would be no unnecessary delay in exceeding Mach 1. According to James Young of the USAF Flight Test Center History Office, Boyd stated that ." . . . the AMC [Air Materiel Command] program would be progressive and it *would* be brief." The colonel knew very well that North American Aviation was nearing the rollout of its new jet fighter and, being an engineer himself, was certain this would be a threat to the X-1. "He knew," Ken Chilstrom recounted, "that if the Germans had likely gone supersonic in a swept-wing 262, or one of their experimental rocket ships [the Natter], then a better, structurally sound jet with a better engine like the XP-86 could certainly do it."

Boyd was not territorial, but there was a lot hanging on this program, both professionally and personally. The Army had spent vast sums of money developing Bell's rocket, specifically to fly faster than sound, so to have a fighter lift off from the ground rather than have to air-drop, then go supersonic with a conventional jet engine would make the whole organization look foolish. In Boyd's defense, he was also thinking about the military test pilot program, the image of the new Air Force, and the future of experimental flight testing.

Contrary to the Hollywood version, there was considerable preparation for this, as in all test programs. In early July, all three military pilots were sent to Buffalo, New York, to meet Larry Bell and get a look at the aircraft itself. While they were there an engine run of the XLR-11 rocket engine was performed and this was an attention getter. Yeager later recollected, "We didn't walk too steady when we left that hangar. . . . That sumbitch scares me to death."

Who could blame him? During the last week in July they traveled out to Muroc to join the X-1, and for an orientation course conducted by Dick Frost, a test pilot himself and Bell's chief flight test engineer. This consisted of long days discussing systems, aerodynamics, flight profiles, and emergency procedures—everything that was known about the rocket ship. At one point Chuck asked if a bailout was possible and Frost simply replied, "No way."

On August 6, 1947, a B-29 piloted by Major Bob Cardenas took off from Muroc and climbed up to 25,000 feet. The bright orange X-1 was released and Yeager, with no engine installed, spent the next eight minutes uneventfully gliding down to Rogers Dry Lake. He did the same thing the next day and also on August 8, during the final glide flight. That same day, barely 100 miles to the southwest, another highly significant aviation event was occurring. The XP-86 prototype was rolled out from Mines Field in Los Angeles, then taken apart and trucked over the San Gabriel Mountains to Muroc. Initial Phase I contractor flight testing was to begin very soon at the hands of North American's chief test pilot: none other than George Welch.

So the final stage was set.

Centuries of aerodynamic theory, decades of flight technology, and years of struggle, war, and hardship for the pilots involved all converged on the roof of the high desert in that summer of 1947. Despite everything that had been in preparation for this, no one *really* knew what existed for a man in the thin air beyond 0.92 Mach. Bullets and artillery shells could do it, and maybe a long-forgotten German pilot had done it, too, yet nothing was for certain except that a demon lived out there. Maybe a demon of science or aerodynamics, but he could also very well be a demon of fame: or of death.

Nine

The Demon

Naked, except for a pair of English riding boots, she cantered a stallion named Dream of Love around a Sacramento competition ring, shouting as she rode; Florence Leontine Lowe *wanted* to be seen and heard—and she was. Born in 1901 to a wealthy and prominent California family, it seemed obvious that she would always be different; one of those individuals who are born out of time, and never quite fit in where they are. There is a choice for people like that. They can either conform and be miserable, or they can try to force the world around them to conform with their life. Florence Lowe chose the latter, almost from the beginning.

Inherited temperament and the turn-of-the-century environment did not help—or maybe it did. Her grandfather, Thaddeus Sobieski Lowe, was an adventurer, avid balloonist, and a restless, creative genius. Using his balloons during the Civil War battles of Chancellorsville and Mechanicsburg (among others), Lowe floated over enemy lines spotting for Union artillery while wearing a black frock coat and silk top hat. Frequently shot at by both sides, Lowe is credited with saving hundreds of Union lives. After the war he patented a number of inventions, including a refrigeration unit suitable for ships or railroads and a process to manufacture ice. His success permitted the construction of a 24,000-square-foot mansion on South Orange Grove Avenue, very near the modern Beverly Hills.

Lowe married and his son, Thaddeus Junior, grew up loving the outdoors and especially horses. Unfortunately, in 1893 a wheat shortage and series of military coups in Argentina panicked investors, who started a run on American banks. The elder Lowe lost it all: the family business, mansion, and investment ventures. However, the younger Lowe had married into an old-money Philadelphia family and was able to maintain the lifestyle of an affluent gentleman, which proved fortunate for his daughter. Thad Junior's mother-in-law built the couple a mansion in San Marino, Pasadena, on South Garfield Avenue and this is where Florence grew up. Riding horses and fishing, she preferred the rough-and-tumble world of boys, which endeared to her to both her father and grandfather, who rarely failed to indulge the child. Her mother believed differently. Young ladies were supposed to be decorous and proper and concern themselves with music, art, and preparation for marriage.

Florence cared for none of these things and rebelled at every chance. "She was a mystery to them," writes Lauren Kessler in *The Happy Bottom Riding Club*, "a little girl who had no interest in being a little girl." Her parents' solution was a series of private boarding schools, four in eight years, to reform the child. Florence knew she was not the ideal girl, whatever that was, and was always keenly aware of her own mother's disapproval.

As she matured into adolescence, she had a choice: to slide into anonymity as a thick-necked, broad-shouldered, round-faced girl and hope some man, sometime, would marry her, or ride though life as she rode the stallion, and not give a damn what most people thought of her.

She chose the latter.

This would be the girl who fled to Tijuana from the Ramona Convent and the Sisters of the Holy Name, who faked suicide to shock her roommate, and who used imported French lingerie to polish her riding boots. She would be Florence Lowe, who pretended to be the governor of California's daughter as she rode naked around the ring in Sacramento. Then, in 1919, she met Calvin Rankin Barnes, a high Episcopalian priest and rector of St. James in Pasadena. He was also a bachelor and quite taken with the lively eighteen-year-old girl. When marriage was suggested, Florence, largely to escape her mother, quickly agreed and the couple married in January 1921.

It did not go well from the start.

Neither could consummate the marriage on the wedding night, and a two-day train ride in separate berths did nothing to improve the situation. Neither did their fourth night together at a resort in the Grand Canyon when Calvin insisted on intimacy. His abrupt, postcoital statement—"I do not like sex. It makes me nervous. I see nothing to it, and do not wish to have any more of it"—closed the book on the subject. Hardly an auspicious beginning, it became more complicated when Florence discovered she was pregnant. A boy was born nine months later, whom she named William Emmert Barnes and then promptly turned over to a nurse. Florence did not like motherhood and had never had a good example of it, nor did she enjoy a life without servants, a mansion, her horses, or money. "She had," Kessler brilliantly writes, "married without love and conceived without passion." So when time came to leave the hospital, Florence went to her parents' mansion, not her husband's rectory.

This began years of a Jekyll and Hyde existence; she discovered sex through a series of lovers and enjoyed the demimonde thrill that accompanied illicit affairs. Florence worked odd jobs during the day, yet taught Sunday school and attempted to play the respectable clerical wife by night.* She tried her hand in Hollywood training horses for the movies and occasionally worked as a stunt double until her mother died in 1923. Her father rapidly overcame his grief and remarried a woman only three years older than his daughter, moving out of town and leaving the San Marino mansion to his daughter. Florence forsook the rectory altogether and resumed life as she had previously known it. Her home, and a Laguna beach house, became wild party spots for the Hollywood elite and those who wished to be. One of them happened to be the future CEO of Western Airlines, and another a young University of California football player named Marion Robert Morrison.†

One day in the spring of 1927, Florence and a few bored friends decided to liven things up a bit by hiring on a banana boat as crew and heading down to South America. The alcohol made it seem a good idea, so Florence dressed like a man and signed on the M/S *Camina* as a "Jacob Crane." The boat sailed, eventually putting into San Blas on Mexico's west coast and

---

* To learn the catechism, she bribed her students with pocketknives.

† Marion Morrison would later assume "John Wayne" as his professional name.

herein lay a problem. Several problems, in fact; first was the ongoing political strife in Mexico, and second was the fact that the *Camina*'s hold was full of guns. After six weeks of confinement, Florence and the helmsman, Roger Chute, escaped. Their plan was to cut across Mexico, get to the coast, and make their way north back into the United States. Chute had a sway-backed nag, but all Florence could find was a burro and she joked that he looked like a modern-day Don Quixote.

"In that case," Chute replied, "you must be his companion, Pancho."

"You mean Sancho . . . Sancho Panza."

Chute laughed. "Ah, what the hell, Pancho or Sancho, you fit the bill. From now on I'm calling you Pancho."

And escape they did. To Mexico City, then Veracruz and eventually up to New Orleans. From there, the pair crossed Texas and made it back to California in November 1927, nearly seven months after leaving. But Florence was far from content. She had finally decided what she wanted: a life of exotic places and dangerous adventures; and a life of men. There was even a new name to match her new identity: she was, and would now remain, Pancho Barnes.

By the end of the first week in August 1947, the Army X-1 team at Muroc was pleased, or at least most of them were. The X-1 had been glided three times and Yeager was getting familiar with the rocket. The NACA group, however, including their test pilot Herb Hoover, was wary of the military; and the Bell representatives, especially test pilot Dick Frost, were unimpressed. Slick Goodlin had done all this over a year before and made powered flights beginning in December. The Army

X-1 did not even have a motor yet and would have to wait a few more weeks due to a shortage of spare parts. In the meantime, the military officers had a visit from Colonel Boyd himself.

He was not a happy man.

Boyd had discovered through Bell and the NACA that Yeager *rolled* the X-1, twice, on his first glide flight. Herb Hoover had said, "This guy Yeager is pretty much of a wild one . . . on the first drop, he did a couple of rolls right after the leaving the B-29!" It did not end there. During Chuck's third glide on August 8, he did a short, two-turn spin and then did a little mock dogfighting with Bob Hoover's chase plane. These antics, in overt violation of the flight test plan, were one reason why military test pilots had not enjoyed a good reputation since the Roaring Twenties and Jimmy Doolittle. Reversing this perception, and altering the reality it had created, was Colonel Albert Boyd's personal quest—and now it was in dire jeopardy at the worst possible time.

Using the interim delay, Boyd arrived at Muroc to see for himself, assess Yeager, and make his feeling perfectly clear. He granted the young pilot's desire to fly past the Mach as soon as possible—that was also the Army's objective—but not by throwing common sense to the winds. If a pilot loses an aircraft and/or kills himself while violating a thoughtfully constructed, clearly stipulated test plan, then there is no way out. It was an unnecessary risk and for such actions there was no defense. A test pilot cannot "cowboy" a plan by making it up as he goes, and that is what Boyd, and the others, knew Yeager had done. He was a fighter jock at heart, not yet a test pilot, and frustrated by the constraints placed on his flying. As Yeager himself admitted, "I'd attend these highly technical NACA preflight planning sessions . . . and not know what the hell they were talking about. But Jack always took me aside and translated the engineer's technical jargon into layman's terms."

Layman?

This was the heart of the problem and Boyd's big concern: Yeager was not a layman; he was supposed to be a test pilot, and that meant possessing a breadth of knowledge he did not yet have. So the colonel did what pilots do when they're frustrated and need to blow off some steam: he went to the bar. And why not? Antelope Valley, California, was hardly Manhattan, Piccadilly Circus, the Place Vendome, or anywhere else these men had seen during the war. Los Angeles was about 100 miles away through the mountains, and the road was primitive. There was a nearby desert watering hole called Ma Greens, close to the base, but whenever Boyd was out at Muroc he headed to his favorite haunt in on the West Coast: the Happy Bottom Riding Club, owned by his good friend Pancho Barnes.*

Life had been a roller coaster for Pancho. Always adventurous and easily bored, she found a permanent solution in the spring of 1928: flying. Introduced to an instructor named Ben Caitlin, Pancho took her first lesson that spring and never looked back. A fifteen-minute lesson cost five dollars and she flew continuously, even buying her own aircraft, a sporty Travel Air biplane. As with all pilots who truly love flying, it was freedom and independence, power, a challenge, and in Pancho's case a way to continuously defy society. "Flying," she once said, "makes me feel like a sex maniac in a whorehouse."

---

* One version for the name's origin, which this author accepts, is that Jimmy Doolittle himself told Pancho that an afternoon on a horse "gave me a happy bottom." There is no doubt, however, that Pancho encouraged the sexual double entendre relative to the hostesses she later employed.

She continued living large, and her parties at Laguna were legendary. In addition to the Hollywood types, she now encouraged pilots of all sorts to come eat, drink, and talk flying. Jimmy Doolittle and Roscoe Turner were among her favorites, and Pancho began flying for oil companies and aircraft manufacturers, like Lockheed, on the side. In an age that still largely underestimated women, it was great publicity to have a female pilot and what better way to prove aviation was safe than to have a woman fly the plane? It was on one of these publicity flights out over the Mojave that Pancho saw green between the big dusty smears of Rogers and Rosamond Dry Lakes. It was just a small alfalfa operation belonging to a rancher named Ben Hannam, but she fell in love with the place and immediately saw that on that sunbaked desert anvil the weather was usually good, and there was always a place to land.

Despite the Great Depression, she was able to continue living large for a bit longer, but much of her inherited wealth was in real estate and the California market bubble had definitely burst. Still, Pancho went through money like it was water and by the early thirties she was spending more in one month than three average families spent in a year; stables, horses, two houses, furs, and now airplanes. It could not last, nor did it. She had violated the only inviolate rule of inherited, invested money: never spend the principal. The beach house at Laguna was gone and she was forced to lease the mansion. Her parents were now both dead, her husband was in New York, and Pancho wanted out.

Remembering the alfalfa ranch, the desert seemed to her, as it has to many others, an escape: a barren, simple place to begin again. In an astonishingly bad business move, but utterly normal for Pancho, she swapped an apartment building on North Sixth and Witmer in downtown Los Angeles for Ben Hannam's eighty-acre alfalfa ranch.* Naming it Rancho Oro Verde, Spanish for "Green Gold," she quickly discovered that at nine dollars per ton, alfalfa would not support her airplanes and Lincolns, so Pancho did other things. She bought cows, fed them the alfalfa to produce milk, and acquired a dairy. With the dairy, she was able to sell the milk to local schools, the Pacific Borax Mining Company, and the small Army Air Corps detachment on the south side of Rogers Dry Lake. Contracting to pick up garbage at Muroc and the Navy base at China Lake, her hired men cooked the garbage and she fed it to her 300 hogs. Pancho then sold the pork back to the military for meat; obviously, some of old Thaddeus Lowe's entrepreneurial spirit lived on in his granddaughter. In 1937 she was doing well enough to begin purchasing land, and by 1940 her little spread now encompassed 360 acres in and around both dry lakes.

She also built an airstrip.

Coming into some money after her grandmother's death, Pancho finally sold the San Marino mansion and devoted herself to her desert "estate," as she saw it. After the war this had evolved into a sprawling, Spanish-style hacienda with a swimming pool in front, a restaurant and bar, guest quarters, stables, and a rodeo arena. Movie stars, celebrities, politicians, and, increasingly, military pilots from nearby Muroc, or "the foreign legion of the Army Air Corps," as it was locally called, were always around. If Pancho liked you, then there were big steaks, lots of booze, swimming, and later, the company of hostesses.

---

* That lot today, a block away from Wilshire Boulevard, would be worth at least $1 million.

Being a famous aviatrix commercially involved with the military and having one of the only places for men to relax meant Pancho knew everything that was going on—even things she was not supposed to know.

Bell's X-1 was finally up again in the California sunshine on August 29, 1947. The Reaction Motors rocket engine had been installed and today was the Army's first powered flight. All pilots flying new aircraft get at least one "Fam flight," a familiarization flight to put all the ground school instruction into perspective. Yeager had glided the X-1, but never used it under power until today. The B-29 lumbered carefully off the runway and climbed up to 25,000 feet with two chase planes in formation. After all the preflight checks were done, Bob Cardenas nosed the big bomber into a shallow dive, and at 255 miles per hour passing 21,000 feet the little orange rocket plane detached.

The flight test called for a series of maneuvers within the known flight envelope defined the previous year by Slick Goodlin. Yeager was to get "a feel" for the aircraft under power, and definitely not exceed 0.82 Mach. Bell's Dick Frost was flying a P-80 in the low chase position when Chuck lit off the first of four rocket chambers. Five seconds later he ignited the Number Two rocket while shutting down Number One, then lit off Number Three after switching off Number Two. According to James O. Young of the Air Force Flight Test Center History Office "to Frost's utter amazement and displeasure [Yeager] deviated from the flight plan and executed a slow roll."

The problem with this, other than intentionally and unprofessionally again violating the test procedure, was that the

XLR-11 had a safety cutoff that automatically shut the motor down if pressure in the liquid oxygen (lox) tank dropped beyond a certain level, which it did at zero g's during Yeager's roll. Everyone watched as Chuck relit the motor, then rocketed upward to the test altitude of 45,000 feet. After Boyd's lecture, and following his unplanned flameout, what happened next was truly astonishing. Yeager shut down the motor, rolled over, and dove for the runway seven miles below him. Radios exploded as the other airborne pilots wanted to know what in the hell he was doing. Again, utterly disregarding the test plan, flight discipline, and good sense, Chuck had decided "to show the brass down there a real airplane," as he later wrote. Leveling off 2,700 feet over Muroc and at 0.8 Mach, Yeager lit off all four chambers again, roared down the runway over lots of wide eyes, then streaked up out of sight. Getting back to 35,000 feet, the rockets ran dry and he then glided down to land.

No one, except Chuck, was happy about this. The NACA team, especially Walt Williams, and Dick Frost from Bell, thought Yeager was undisciplined. Frost, as a test pilot himself, was thoroughly displeased and chewed the young officer out. Even Jack Ridley, close as he was to Chuck, could not explain away his friend's recklessness and disregard for test procedures. "He [Yeager] didn't have a lot of test experience," Ken Chilstrom remembers. "But Al Boyd thought that didn't really matter for what had to be done with the X-1. We just needed someone to ride it through the so-called sound barrier, and Yeager could do that all right."

Boyd was once again an unhappy man.

He had previously come all the way out to Muroc to give his pilot a face-to-face lecture *and* another chance—then Yeager pulled a stunt like this. The colonel insisted on an official ex-

planation, and Chuck actually asked Jack Ridley to write one
out: Ridley refused. Yeager later wrote about the phone call he
received from Boyd. "Damn it"—the colonel was angry and
concerned—"I expect you to stick to the program and do what
you are supposed to. Don't get overeager and cocky. Do you
want to jeopardize the first Air Corps research project?" Chuck
answered he did not, of course, to which Boyd replied tersely,
"Well, then obey the goddam rules." Yeager tried to explain
that he violated the plan and Boyd's direct orders because he
was excited. The colonel might have bought this, or at least understood it coming from a young fighter pilot, because he did
not ground Yeager or remove him from the project, which is
what most of the others expected.

But if Yeager had been replaced, then the chief of Flight Test
would look foolish for having picked him over more experienced test pilots. It is possible that Yeager was chosen because
he was so junior and if, God forbid, something happened, then
the pilot would get the blame and not the Flight Test Division,
though Boyd's judgment would be questioned. This could con-
ceivably put the colonel's plans for a true military flight test program at risk and, equally concerning, were the politics of the
situation. If the program suffered any mishap, then the Army
Air Force would lose face and Stuart Symington, inaugural secretary of the new Air Force, would certainly not tolerate this or
forgive the man who allowed it to happen. The military could
even lose the X-1 contract, and responsibility for chasing the demon would fall back to Bell, or the NACA, or be canceled altogether. If that occurred, the United States would lose credibility
and its hard-won, albeit temporary, tactical advantage enjoyed
over the Soviet Union.

Then there was North American Aviation. The entire Flight

Test Division was aware of the XP-86's potential, but from Boyd's point of view it was essential that the X-1 get through the Mach before the Sabre. If a jet fighter could take off convention- ally and exceed Mach 1, even in a dive, then several years and millions of dollars would have been wasted. North American would get the credit, not the Air Force, and that would cause Al Boyd serious career problems. The Sabre was also at Muroc and had completed its high-speed taxi tests the same day Yeager flew the rocket, so time was very short. No, it had to be the X-1, and without the delay incurred by replacing the pilot so, in effect, North American's XP-86 saved Yeager from obscurity.

During September the X-1 flew four additional powered flights and Chuck, mindful of Colonel Boyd's warning, at- tempted to follow the flight test plans. The NACA wanted tran- sonic stability data from an actual aircraft, not a wind tunnel, for the 0.80 to 0.85 Mach regime. This was provided, and buf- feting accompanied by a wing drop was consistently recorded, although Yeager was able to maintain control. By September 10, his fourth powered flight, Chuck was able to get past 0.9 Mach for the first time, and two days later the X-1 hit 0.945 Mach, which provided the engineers a superb opportunity to assess stabilizer and elevator effectiveness. Another wing drop occurred, but at this speed it was accompanied by *severe* buffet so Yeager cut the power and glided back to Muroc.

What happened was as fascinating as it was dangerous, but once again proved that wind tunnel tests and unmanned air- craft were no substitute for a test pilot. The shock wave had, as predicted, moved aft as the Mach number increased and was hovering over the horizontal stabilizer, directly over the hinge line, unfortunately. Like the Me 262, the X-1 horizontal stabilizer could be manually adjusted via a trim switch on the control column, which was set to move at 1 degree per second, and that was fast enough. The solution: a faster actuator that corrected at 3 degrees per second, and on future flights the pilot would use normal controls until elevator effectiveness was lost, then use the trim switch to move the stabilizer and regain some pitch control. Yeager's ability to keep control at all confirmed that raising the tail, just like the Germans had done, was effective in keeping the control surfaces generally clear of transonic burble. On September 18, 1947, six days after this flight and while the faster actuator was installed, the National Security Act took effect, and the Army Air Forces officially became the United States Air Force.

On Monday morning, September 29, 1947, George Welch walked out on Muroc's flight line toward a jet waiting in the sun. North American's elegant XP-86 had been reassembled two weeks earlier but, during an engine run, an unwary crew chief had been sucked into the intake and killed. The jet needed a new engine and had to be thoroughly checked before testing could proceed, but that was done and the Sabre was now ready for the its high-speed taxi test. This was normal, a chance to check out all the systems, controls, and engine without getting fully airborne. The sleek, silver prototype was gleaming in the early light, and the idea was to finish by midmorning before the dry lake really got hot. Wearing a blue flight suit and a bright orange Lombard helmet, Welch used an external power cart to start the General Electric J35 jet engine, then taxied out to begin testing.

The jet was beautiful and the cockpit was a pilot's dream.

Both side consoles contained noncritical controls like the radio and temperature switches while everything he needed to fly the fighter was right in front of him. Forward of the stick the main panel was T-shaped, with the primary control instruments in the center, and the engine gauges in plain sight along the top. Worlds away from the P-40 and P-38 cockpits he had lived in just a few years earlier. The throttle was actually an upright grip mounted on quadrant by his left knee, and he slid it forward, accelerating multiple down the dry lake. Welch would let the nose wheel lift off then bring the power back to idle, checking and rechecking all the systems. One objective was to verify that the recording oscillograph and cameras were functioning; another was to fine-tune the horizontal stabilizer setting for takeoff. Welch did all this and taxied back in by ten o'clock to debrief, confer with the engineers, and write his report. Based on all this, the NAA team could decide when the first flight would be scheduled, and in the meantime George would remain at Muroc.

Due to the accident with the crew chief, he had spent most of the month working two other projects: the P-82 Twin Mustang, and accomplishing spin demonstrations for the XSN2J-1, a new Navy trainer. Something had happened in connection with the Navy plane that intrigued him greatly. While ferrying the plane back east to Patuxent River, Welch stopped in Deming, New Mexico, to spend the night. At a local bar he was invited by some locals to participate in what passed for entertainment in Deming, driving out to the San Andres Mountains near Las Cruces and watching rocket launches. Captured German V-2 rockets were fired from Las Cruces northward over the mountains into the White Sands area and, about thirty seconds after launch, a huge shock wave was felt. The locals got a kick from the rush of pressure and, more to the point, thought the thunderclap *ba-boom* was exciting.

Though he did not go with them, George apparently perked up at the mention of noise. It was well known that the V-2 could go supersonic, so the noise had to be what Theodore von Kármán long predicted: that the shock wave from a supersonic object, not just a wing but the whole aircraft, would generate a wave that rolled out and down and be audible to those on the ground. Now, late in the month, Welch was back and would remain at Muroc until the tests were complete, and he was still thinking about the noise.

In a little British MG, he could make it through the San Gabriels by way of Soledad Canyon in ninety minutes from his home in Brentwood, and he usually stayed in one of the twenty guest cabins at the Happy Bottom Riding Club. In Pancho's bar that night, he heard about the X-1 trim issues and knew they were close to fixing it. He also knew Yeager and was aware that the Air Force pilot had gone past Mach 0.92 and would very likely try to go supersonic in the next few weeks. A thinker and engineer as well as a pilot, Welch spent a great deal of time with the NAA designers: Larry Greene, the chief of aerodynamics; Ed Horkey, the performance engineer; and Walt Williams. They all agreed that if the swept-wing XP-86, even with the J35 engine, began at least a 30-degree dive from 35,000, then the jet would accelerate through the speed of sound. They believed this had also been possible for the Me 262, though there was no record of it, and the Sabre was much, much stronger. Tony Weissenberger's structural design was extremely good; the wing was tough and stiff without a lot of extra weight so it could handle the stress. George Welch and the XP-86 were scheduled for their first flight on Monday, October 1, and according to one source, George had said, "I don't know what Yeager's waiting for, but I'm ready."

Besides the highballs, scotch, and barbecues, the Happy Bottom Riding Club was famous (or infamous) for the hostesses who worked in the lounge. "My girls are sugar to catch the boys," Pancho once said, and meant it. Periodically returning to Los Angeles, she interviewed models or actresses who were down on their luck in Hollywood, then selected the most attractive girls to work in Antelope Valley. Rumor was each girl was given a "Smith" surname, and a day of the week or a month as a first name. Regardless of the talk, she did have house rules posted and among them were:

1. We're not responsible for the bustling and hustling that may go on here. Lots of people bustle, and some hustle. But that's their business, and a very old one.
2. We permit you to accept tips and gifts while performing your regular duties. But NEVER are you to accept money in remuneration for the more intimate aspects of romance.

Wives—usually they were military wives because the civilians were there alone or drove in from Los Angeles—took a dim view of the Happy Bottom Club. Glennis Yeager called it "a desert whorehouse," and likely gave her husband a great deal of grief about the time he spent there. A local school principal apparently spent early Monday mornings walking the road with a stick to spear all the used condoms before the kids rode by on their bus. Chuck Yeager wrote that "her [Pancho Barnes's] place was a big part of the sixteen years I spent out on the Mojave . . .

she would never use a five- or six-letter word when a four-letter word would do. Flying and hell-raising . . . one fueled the other. And that's what Pancho's was all about."

No question the place was wild and raunchy, but why not? These were men who had lost their childhood to the Depression, their young manhood to the Second World War, and were generally married to women who did not understand them. Only combat veterans can really talk to other combat veterans, and fighter pilots to fighter pilots. Everyone else really doesn't get it, and never can. Add to it that these guys were under enormous stress, chasing the demon into the unknown with no real assurance they would live to tell about it. Their excesses, real or imagined, are not for us today to judge.

And it wasn't just the young captains and majors who were interested in the place. Colonel Al Boyd loved Pancho's, and her barbecues and pool parties. One day a pair of generals from Wright Field called the club to tell her they were coming in for the weekend. They wanted steak, scotch, and a little female company. Pancho had her cooks make huge sandwiches—with two naked girls as the stuffing—and her ranch hands carried the platters out to the generals. It was a world far removed from what most of these officers endured every day, when the threat of another world war was becoming quite real and diversion was much needed from the pressures of a new military service, a new Red Scare, and a perilously fragile political situation.

George Welch, after all he had been through, was no different.

He had a hostess friend named Millie Palmer and, according to one source, had dinner with her at Pancho's the evening of his successful high-speed taxi tests. Millie was a very good source of information about the X-1 program, not that there were any secrets kept at Muroc anyway. Her cousin was a flight engineer aboard the C-54 Skymaster that handled logistics for the X-1 program. The cargo plane was based out of San Francisco and a frequent VIP aboard was none other than Major General "Jumpin' Joe" Swing, who planned the airborne phase of Operation Husky four years earlier. A classmate of Dwight Eisenhower and Omar Bradley, Swing was extremely well connected and widely admired for his combat prowess. He had chased Pancho Villa into Mexico in 1916, then fought in France with the U.S. 1st Division during the Great War.

After Operation Husky, Swing led the 11th Airborne through the invasion of the Philippines and flew his own reconnaissance aircraft around the island during the fighting. He also parachuted into Luzon with his men and freed 2,134 American prisoners of war from the notorious Japanese Los Banos prison. Swing and his men killed all the Japanese and saved every POW; he was awarded the Distinguished Service Cross to accompany his three Silver Stars and, rare for an Army paratrooper, two Air Medals. In addition to his courage and athleticism, Swing was extremely intelligent (he graduated higher than Eisenhower) and thoroughly understood both aviation and aerodynamics.

A great friend of Pancho's and Welch's, Jumpin' Joe was a true soldier who despised prima donnas and subterfuge, though he understood the political realities during the fall of 1947. The general liked to get out to the Happy Bottom Riding Club whenever he could, ostensibly to ride horses as he had been an accomplished polo player at West Point, but the scotch and hostesses no doubt added to his attraction for the place. Fortunately, the C-54 made the trip often; San Francisco to Muroc to Buffalo and back—as many times as necessary, filled with equipment, parts, and people who talked too much. Through Millie Palmer and the crowd at Pancho's, Welch would have been well aware that the X-1's pitch control issue was nearly solved, and that the Air Force was pushing hard. Yeager would certainly attempt Mach 1 within weeks, if not days.

George was also quite intelligent and he understood aerodynamics, both in theory and practice. He was certain that the frequent thunderclaps heard around Las Cruces were caused by supersonic V-2 rockets; it fit with what experts like von Kármán long predicted. North American had constructed a very strong, swept-wing, jet-powered aircraft that was, according to Larry Greene and Walt Williams, quite capable of exceeding the speed of sound in a dive. If he did it, and given his temperament he was certainly considering it, Welch knew it would have to be done unofficially. North American, specifically Dutch Kindelberger, had been good to him and he had no intention of causing a problem by incurring the wrath of the U.S. Air Force. The military paid the bills, and the Sabre contract was worth hundreds of millions of dollars.

It was also easy to figure out why the new USAF needed the publicity of breaking the so-called sound barrier before anyone else, especially a civilian company. In fact, Kindelberger had been warned in September by Secretary Symington that Bell and the Air Force would be first, no matter what. Larry Bell had heard the rumors about the XP-86's capability and was apparently furious; he knew very well what another failure would mean for his company, and he used his considerable political influence through President Harry Truman to pressure NAA into laying off.

Dutch Kindelberger reportedly said, "I don't give a happy damn," about who got official credit for busting the Mach.

He was a businessman and as long as NAA got the contracts, someone else, even Larry Bell, could take the credit. This attitude paralleled George Welch's perfectly: he wanted to do it because glimpses of the demon's tail were not enough; he wanted to catch the thing, or at least chase it a bit farther out. Supersonic flight was the next big challenge, and Welch existed for challenges of any type. The official credit was of no more consequence to him than recording his kills in combat. One retelling of that Monday night at Pancho's has George Welch asking his friend Millie Palmer to be his data recorder. To listen up on Wednesday morning for a "sharp boom like a clap of thunder," and to write down the time and the reactions around her.

Whether this happened or not, George Welch did take XP-86 PU 597 off the ground from Muroc early Monday morning, October 1, 1947—exactly four years after the flight of America's first jet, the XP-59. He headed south, enjoying the smooth, powerful plane and climbing up easily into clear skies with fellow test pilot Bob Chilton flying chase with a F-82 Twin Mustang.* It was Chilton who told him that his main gear doors were not shut, and after cycling the handle, George got all three wheels up and locked. The P-82 remained over El Mirage Dry Lake, east of Muroc, and Welch spent the next ten minutes getting the Sabre up to 35,000 feet. Because this was just a familiarization flight, he was to get a general feel for the jet's flying characteristics and check out all the systems. There was no recording instrumentation aboard, very likely the only flight

---

* The new USAF serial number was 45-59597. The old Army Air Force "P" designation for fighter aircraft was replaced with "F" after September 18, 1947.

where this would be the case, so if he intended to push past the Mach it had to be now. Eyes darting around the cockpit, George let the airspeed jump to 320 knots then he rolled left, the stick hard against his leg, and dropped the nose into a 40-degree dive. Leaning forward, he stared through the clear canopy and pointed directly at the only green area visible: Pancho's Happy Bottom Riding Club.

According to the NAA flight test report filed by Welch, as he passed 29,000 feet the wing rolled slightly and the airspeed indicator jumped from 350 to 410 knots. Steepening the dive to 45 degrees increased the airspeed to 450 knots. Leveling off at 25,000 feet, he slowed the Sabre down, checked everything, then did two big barrel rolls to slow further. Meeting up with Chilton over El Mirage, the pair descended into Muroc where Welch again had trouble. This time the nose gear would not fully extend until the jet slowed below 77 knots on the runway, when it finally locked into place. Later that day George Welch left for Los Angeles to debrief with the project engineers, but apparently he called Millie Palmer before departing. She told him that the sound was exactly how he described and it shook the walls. Millie, according to Al Blackburn in *Aces High*, also said, "Pancho is really pissed. You know how nuts she is about Yeager."

There are some problems with this account. First is veracity; the story is hearsay, though the sources are credible. The second problem is scientific. Recall the effects of air temperature and air pressure on the speed of sound. As air gets farther from Earth it becomes thinner and colder, thus less velocity is required to exceed the speed of sound. Temperature above a few thousand feet decreases at a fairly constant 5.5 degrees per thousand feet, so at Welch's 30,000-foot altitude this would be

about −48 degrees Fahrenheit.* Mach 1 here would be approximately 678 mph, or 589 knots. According to Welch's data at
30,000 feet, he saw 450 knots, or 519 mph, which equates to
0.76 Mach: far too slow.

However, instrument gauge inaccuracies, particularly airspeed indicators that rely on ram air pressure, were common
during high-speed flights. Ram air pressure becomes extremely unpredictable at higher speeds, thus often producing
wild, unpredictable readings, so it is certainly possible that the
Sabre was going much faster than indicated. The other indica-
tions Welch reported, like a wing dropping in conjunction with
an airspeed jump, indicate this was the case. Mach effect, as
it is called, results from shock wave formation and radical airflow changes within the transonic region. Recall also that the
high transonic X-1 flight profiles yielded similar results, and at
a higher indicated Mach number, so it is reasonable to conclude
the XP-86 suffered the same effect.

Many of the patrons at Pancho's were aware of the two flight
programs, and some understood enough to know what they
were hearing. Millie Palmer was not alone in this and, though
Pancho Barnes tried to blame the noise on nearby mining operations, no one had ever heard explosions coming from the
mine before. Sonic booms are very directional: they propagate
in the direction of the supersonic object's travel, which, in this
case, was west from El Mirage toward the base. Unless someone was directly in line, or close to it, the sound would be much
less pronounced, more like a dull, distant boom that could be
mistaken for an explosion.

---

* Table of U.S. Standard Atmospheric Heights and Temperature.

So what is fact is that on that October morning George Welch had everything he needed to chase the demon. That is, the correct airframe with the necessary swept wings; the right engine; and without question the right skills and attitude. He also had the correct parameters: at least 35,000 feet in altitude, 350 knots of airspeed, and nothing but flat, dry desert below. In level flight the J35 jet engine could never have produced the thrust to bust the Mach but in a dive it, and the XP-86, could push past the speed of sound—if Welch did it. The rest is conjecture, but it is unemotional, informed, and agendaless conjecture. If Welch's instrument readings, which are consistent with high transonic speeds, are taken with the nature of the man and testimony of those who heard "something" that morning, it is extremely likely that the XP-86 achieved supersonic flight on October 1, 1947.

The next few days were as eventful as they are largely unknown. On October 3, the X-1 flew its sixth flight to test the new trim settings for the horizontal tail actuator. The NACA team had also been noticing the Mach effect, as Yeager routinely reported lower indicated numbers than they were recovering in the postflight data. For instance, on this sixth flight he showed 0.88 Mach, but the NACA folks read 0.92 from the instrumentation package. With the benefit of this knowledge, it is more certain the XP-86 was flying faster than George Welch observed in the cockpit. This phenomena was repeated on the seventh rocket flight, which Yeager recorded as 0.92, but was truly 0.945 Mach.

Welch flew again on October 9, this time with the Sabre's gear locked down for low speed stability and control checks.

Landing back at Muroc after his second flight, he was told his wife, Jan, was in premature labor. Grabbing the NAA logistical aircraft, he flew to Santa Monica and was in her St. John's hospital room by early afternoon. While this was happening, the X-1 was also airborne on its eighth hop, and at 0.94 Mach suffered a complete loss of elevator effectiveness—no pitch control at all. Yeager instantly shut the rockets down, jettisoned the fuel, and glided back to Muroc. Proving his practical brilliance yet again, Jack Ridley calculated that the entire horizontal tail could be manipulated, in increments of a degree, at high Mach numbers and this would provide enough pitch control for safety. "It will keep you flying," he said, and Yeager believed him. "I trusted Jack with my life," he admitted quite honestly.

That evening it was discovered that Yeager had actually achieved 0.997 Mach, rather than the 0.94 he read from the gauge. He was close, he knew it, and he wanted to chase the demon all the way. One account states Pancho told him that night about Welch's boom nearly two weeks earlier, and that Yeager dismissed it as a fluke—whatever that meant. General Joe Swing had flown in on the C-54 for the weekend and was also at the Happy Bottom that night. He had been hoping to see George Welch, whom he knew well from his time in Australia during the war and had kept up with during his frequent trips to Muroc. Swing also knew Stuart Symington, the new secretary of the Air Force, and the general was discussing the real meaning behind the X-1 program with the NAA folks. According to Al Blackburn, the paratrooper called Yeager a "creative thespian" and laid out the political implications of anyone other than the military breaking the "wall."

Over the weekend, George Welch brought his wife home to Brentwood, but Giles, his three-pound, premature baby boy, would have to wait. The next XP-86 flights were to be made with the gear down to complete the low-speed tests, and apparently George was suspicious of this. The timing was too convenient, and it seemed likely that someone at Muroc had heard his October 1 sonic boom, put it all together, and called Washington. With the gear extended there was no way to break the Mach again, so Welch asked for, and was given, permission to not *lock* the gear on the ground, but simply keep the wheels down in flight unless safety dictated otherwise. In any case, his flights were scheduled for Tuesday, October 14: the same day the X-1 would fly again.

But Chuck Yeager also had problems. On Sunday night he and Glennis ate dinner at Pancho's then decided to go for a night horseback ride. Coming back to the Happy Bottom, the pilot failed to notice someone had closed the gate to the corral and he and the horse hit it dead-on. Yeager flew off and was "knocked silly," as he put it. Having trouble breathing, the following day Chuck and Glennis went to a civilian doctor in nearby Rosamond, who taped up his three broken ribs. Worried that he could not shut the X-1's peculiar hatch, Jack Ridley improvised a ten-inch wooden handle for Yeager to use.

Tuesday, October 14, 1947, dawned clear and quiet. George Welch had arrived very early from Los Angeles and was airborne after sunrise with Bob Chilton again flying chase. The Sabre's sixth flight lasted exactly six minutes; the airspeed indicator had not been reconnected properly. Refueled and airborne again, Welch spent thirty minutes or so climbing up to altitude and running through the test card, reading off results as he went. At some point, apparently with Chilton's prearranged concurrence, he retracted the gear and disappeared.

By this time, B-29 "Eight-Zero-Zero" had gotten airborne

off runway 06 from Muroc with the little orange rocket hanging off its belly. Dick Frost was flying low chase and Bob Hoover was climbing up to the high position at 40,000 feet some ten miles ahead of the bomber. Muroc Tower directed all aircraft to stay clear of the dry lake, and at 1025 Jack Ridley asked Yeager if he was ready to go and told him to remember the stabilizer settings. "Hell yes . . . let's get it over with," was the reply. From 20,000 feet and indicating 250 miles per hour, the X-1 dropped free at 1026, just as the silver XP-86 landed at Muroc.

The actual transcript, courtesy of Dr. Alex Spencer from the Smithsonian Air & Space Museum reads:

YEAGER: Firing Four . . . Four fired okay . . . will fire Two . . . Two on . . . will cut off Four . . . Four off . . . will fire Three . . . Three burning now . . . will shut off Two and fire One . . . One on . . . will fire Two again . . . .Two on . . . will fire Four . . .

RIDLEY: How much of a drop?

YEAGER: About forty psi . . . Three on , , , acceleration good . . . have had mild buffet . . . usual instability. Say, Ridley, make a note here. Elevator effectiveness regained.

RIDLEY: Roger. Noted.

YEAGER: Ridley! Make another note. There's something wrong with this Machmeter. It's gone screwy!

RIDLEY: If it is, we'll fix it. Personally, I think you're seeing things.

YEAGER: I guess I am Jack . . . will shut down again . . . am shutting off . . . shut off . . . still going upstairs like a bat . . . have jettisoned fuel and lox . . . about thirty percent of each remaining . . . still going up . . . have shut off now.

Reaching Mach 1.05 at 42,000 in the vicinity of Victorville, Yeager toggled off all four chambers, which cut off the fuel and lox, then pulled back on the yoke with both hands. Trading all that airspeed for altitude, Chuck and the little aircraft put the earth behind the tail and soared upward; a speck of orange slicing through the blue so far away from the tan desert and its dirty white lakebeds miles below. There was no horizon, just sunlight and the clear air of a vast empty sky . . . and the demon. With no power, the X-1 gradually slowed and at 45,000 feet, over seven miles above the Mojave, it shuddered into a 1-g stall.

With both hands wrapped around the yoke, Chuck watched cockpit dust float up as the nose dropped and he got light in the seat. After a long, quiet glide Yeager landed smoothly on the dry lake at 1040: a mere fourteen minutes after his release. There was then, as now, no physical drama in supersonic flight, no black hole and no wall in the sky. No demon; at least, no demon that could be seen or felt. It had already fled deeper into the thin air and was waiting for whomever would come next. Chuck was disappointed with the simplicity of passing Mach 1 and recalled, "It took a damn instrument meter to tell me what I'd done." His brief, six-paragraph report also stated "stability about all three axes was good as speed increased." There were no issues and the whole thing was anticlimactic. A "poke through Jell-O," as Yeager remembers his 20 seconds of supersonic flight.

Back at Muroc the engineers verified that the X-1 had indeed reached Mach 1.06, about 700 miles per hour at 43,000 feet, officially flying past the speed of sound. Everyone wanted a party, but Colonel Al Boyd locked the situation down so the military and Bell crowd went to Yeager's house, where, understandably, they got plowed. Chuck got drunk, rode off in the dark on a motorcycle, and Ridley found him sprawled on his back in the road. George Welch, under no such official ban, was at the Happy Bottom Club with Major General Swing, who asked him about the sonic boom and the cracks in the Pancho's east-facing windows. The test pilot, who loved pranks and gags, is reputed to have innocently replied, "Maybe a V-2 flew off course out of White Sands." Swing apparently laughed—he knew what had happened.

Then he asked why there were two booms . . . about twenty minutes apart.

So the deed had been done.

The engineers analyzed their reams of data; the Air Force kept a lid on the big secret and the pilots basically shrugged and went back to work on other things. Without detracting from the physical feat itself, it should be apparent that Jack Woolams, Slick Goodlin, or really any fighter test pilot (like Jack Ridley) could have flown past the speed of sound; Goodlin certainly could have done this anytime in 1946 after the X-1 got its rockets, but he did not because he was a professional. Ken Chilstrom was unequivocal about the situation. "Slick Goodlin was a much better test pilot than Yeager," he said. "No contest." Exceeding Mach 1 was not part of the Bell contract and Chal, as his friends called him, was not a cowboy.

Chuck Yeager was.

And that's not necessarily a bad thing. Yeager, according to those who knew him, was exactly what you would want a World War II combat fighter pilot to be: supremely confident and physically skilled with fast reflexes. Exceedingly calm under pressure, he was convinced enough of his own immortality that calculated risks did not bother him a bit—he figured he could get himself out of any situation he got himself into. These are defining characteristics of successful fighter pilots and, hopefully, always will be.

But a test pilot is something different.

Granted, the late 1940s were a transition period from the ballsy stick-and-rudder types to those who actually understood and could analyze the aerodynamic phenomena they were witnessing. This is precisely why Glen Edwards, Gus Lundquist, and others were sent back to top-notch schools to complete advanced engineering degrees. Times were changing and, as the future astronaut program would make quite clear, just being a great pilot was no longer enough for this type of flying, which is precisely why Jack Ridley was an essential, deal-breaking part of the X-1 team. Yeager and Hoover were good 'ol boys; any competent pilot could have flipped the switches and rode the rocket past the Mach—it just happened to be Yeager. And he did do it, which should always be remembered. He knew the risks, and he knew that in spite of the conviction that Mach 1 was just a number, no one absolutely *knew* for sure.

Except perhaps George Welch.

This author's *personal* opinion, based on the men, aerodynamics, aircraft, the timing, and my own gut judgment as an ex-fighter pilot, is that George Welch caught the demon first. He did the same Mach run during his first flight in the F-100 six years later so why, with the aircraft and opportunity, would he have not done it in 1947? Yet General Yeager, who has been hounded by this question for decades, has an excellent point: Where is the data? Can anyone prove Welch and the XP-86 exceeded Mach 1 before Yeager did it officially on October 14,

1947? Not yet. But if there is hard evidence out there, then hopefully it will eventually surface, and in the meantime the converse is equally valid. Can anyone prove that Welch did *not* bust the Mach early in October 1947, over the high desert? Other pilots, including Ken Chilstrom who flew that exact Sabre, state that from 35,000 feet, in a full power dive, it would positively go supersonic. The jet was strong enough to withstand transonic forces and, unlike the Natter or Me 262, could withstand the recovery to subsonic flight.

In *either* case, it was done.

If what occurred that morning was the truth, then it shall endure; if it was not, then someday the truth will emerge. Truth always does. What *is* certain is that on an October day in 1947 a man officially flew past the speed of sound and buried the myth of a sound "barrier" forever. There were other barriers to break now, some very real, some exaggerated, and some as yet unknown. Perhaps this is the century where man chases the demon into the thin air as far as he can go, to the limits of our capability, science, and imagination. Perhaps it is not. Maybe the demon cannot be caught but only exists to lead us ever deeper into the unknown and, in the chase itself, it continues to teach us that mankind truly has no limits.

Epilogue
RIPPLES

t is formally labeled a capillary wave, "a traveling wave affected by surface tension," though most of us would simply call it a "ripple." Ripples occur each day for every human in big or small ways; every action taken, each choice made has consequences. Often not noteworthy, they are usually ignored altogether or absorbed into what each of us calls life. Yet some events do occur, manmade or natural, that create visible ripples and alter the very fabric of our existence. What if, for instance, an anonymous Austrian farm girl named Anna Maria Shicklgruber never gave birth to an illegitimate child she named Alois who, in turn, fathered Adolf Hitler? Would the Second World War have occurred and over sixty million humans, at least 3 percent of the total population, lose their lives? Very likely, but certainly not as it did, and would the motivation then have existed to develop the rocket and jet which resulted in manned flight past the speed of sound?

Ripples.

Immediate public consequences for breaking the so-called barrier were negligible because the public was unaware the flight had even occurred. The military did what was usual when undecided about what to do, and clamped the security lid down tight. This hardly meant what had happened was a secret since everyone at Muroc either heard the boom (or booms) and listened to the talk. As for the public, nothing would be confirmed until *Aviation Week* published an uncleared article about the X-1 on December 20, 1947; an article that did not go over well at all with the secretary of the Air Force who threatened to prosecute the editor, Bob Wood, for disclosing official secrets. Dutch Kindelberger of North American Aviation was at the Pentagon directly after the article released and was summoned to Stuart Symington's office. Apparently the secretary, though upset over the disclosure, was more concerned about a phone call he had received from an old acquaintance: Joe Swing.

The general relayed to the secretary that the article was false, that the X-1 had not been the first aircraft to go supersonic and Yeager was not the first to do it. Of course, Swing had heard the rumors about the first flight of the XF-86, but had also been present at Muroc himself on October 14 when both aircraft were airborne, and he specifically heard two booms fifteen or twenty minutes apart. This was no hostess or uninformed mink rancher, this was a general officer and a war hero. Symington, who owed his position to fellow Missourian Harry Truman, had guaranteed the president that the U.S. Air Force, and Bell's X-1, would be the first to fly supersonic and now there was credible evidence to the contrary.

So an agreement was reached that hopefully satisfied everyone. If any supersonic XF-86 documentation existed, then it would be contained by the company, as would any public verbal discussions. "We just don't talk about it," NAA test pilot Bud Poage said. "Not then. Not now." The cloistered documentation was to include data from radar-phototheodolite equipment the NACA used to track the X-1 and, according to physicist Dr. Joe Baugher, which recorded the XP-86 on October 19 and 21 at Mach 1.02 through 1.04, respectively. This same equipment was also used to *officially* track the XF-86 on November 13, 1947, when George Welch was clocked at Mach 1.02 diving the Sabre at Rogers Dry Lake. These "High Mach Number Investigation" flights, conducted between October 14 and the onset of military Phase II tests in early December, are quite interesting. The NAA test reports repeatedly list the test Mach number achieved as "over .90," nothing more exact. This is odd because all the other tests show very specific readings with the Mach number often carried out to three digits. Of course, the ambiguity could be due to the before mentioned instrumentation inaccuracies, though these errors were on the cockpit gauges only, and not from the engineering data obtained in postflight analysis.

Nothing was said about this at the time because the X-1's flight had not yet been acknowledged, and after it had been the point was hopefully moot from the Air Force's perspective. Ken Chilstrom, who was the first military pilot to fly the Sabre, took over Phase II testing with the same prototype in early December. "Colonel Boyd told me not to fly past Mach 0.9," he recalled. "That's what North American had officially designated as the jet's top speed. So I didn't fly any faster than that." The company's Preliminary High Speed Performance table shows a bit more. In fact, Welch's fifteenth through twenty-seventh flights chart speeds past 0.9 Mach and, strangely, data for several of these flights is not plotted at all. Of course, there could be a valid reason for this, but it is not noted. However, there is a note on the bottom of the October 1947 reports that reads:

THIS FIGURE SHOWS THOSE DATA PRESENTABLE IN ACCORDANCE WITH THE MAXIMUM MACH 1.0 LIMIT. COMPLETE DATA FOR THESE FLIGHTS WILL BE FORWARDED IN THE IMMEDIATE FUTURE UN-DER A DIFFERENT CLASSIFICATION.

One interpretation of this could be that North American was well aware the Sabre could go supersonic in a dive and was hedging its bets against the future. Could it be done? I asked Colonel Chilstrom point-blank during an interview, and he merely smiled. Even after all these years, his loyalty to Al Boyd is still strong. "He was my mentor," Ken added. "And my friend."

Whether or not the conversation between Symington and Kindelberger actually occurred, there are several facts worth considering. North American did indeed keep its F-86 con-tract, and the initial purchase order of 221 aircraft remained in place. Then the USAF rather lamely acknowledged in June 1948 that George Welch had *officially* gone supersonic in a shallow dive on April 25, 1948, in the XF-86 with the less powerful J35 Allison jet engine. No doubt this announcement was precipitated by British test pilot Roland "Bee" Beaumont's supersonic Sabre flight in late May just prior to the Air Force's acceptance of its first production F-86A models. In the fall of that year, Major Bob Johnson would set a new Federation Aéronautique Internationale (FAI) speed record of 670.981 in Sabre 47-608. Incidentally, the FAI *did not recognize* Yeager's X-1 supersonic flight as a legitimate record because the aircraft did not take off from the ground under its own power, as did the Sabre.

Second, the value of NAA contracts exceeded $300 million within a few years of the X-1's flight, and the XF-86 did indeed become the F-86, a hugely successful jet aircraft from both a tactical and financial viewpoint.* North American would eventually produce over 6,600 of these in the United States, and at least 2,500 under foreign licenses at a flyaway cost of approximately $311,000 each. Another 1,146 FJ Furys, essentially a naval version of the Sabre, were built for the Navy and Marine Corps. The company also held long-running contracts to produce training aircraft, including some 15,000 T-6 Texans, and another 2,000 T-28 Trojans. Interestingly, NAA had acquired enough supersonic data from *someplace* to embark on a series of evolutionary X-plane designs, including the amazing X-15, which was designed for hypersonic flight and eventually reached a maximum recorded airspeed of 4,520 miles per hour.†

Bell Aircraft would meet a different fate. Touting its expertise with rockets, the company was awarded a 1946 contract to build a supersonic air-to-surface missile capable of challenging the Soviets. An offshoot of several programs, it eventually became the ASM-A-2/B-63 and was topped with a nuclear warhead: America's first such weapon. The military redesignated it as a GAM-63 RASCAL, and the Air Force aimed to arm its new B-47 jet bombers with an air-launched version,

---

* Nearly $3 billion worth of business in 2018 dollars.

† Hypersonic speed is above Mach 5 (3,836.35 mph at sea level) while less than 56 miles (90 km) above Earth.

though this was eventually scrapped. Bell then attempted
to develop several X-planes of its own, but never quite succeeded. Unable to compete, the company got out of fixed-wing
aircraft and became one of the world's leading helicopter manufacturers.

The long-term ripples from October 1947 still endure
through today. For the military, exceeding the speed of sound
meant another fundamental change in capabilities that disturbed the muddy waters of diplomacy and foreign policy, just
as the biplane did over the trenches during the Great War, and
heavy bombers did in World War II. Rockets, largely based on
German wartime technology, enabled supersonic delivery of
nuclear warheads and were now a fearsome reality that had to
be countered. Mass-produced, reliable jet aircraft meant high-
speed, high-altitude penetrations of enemy airspace and, when
coupled with nuclear weapons, drastically altered military
thinking around the world, especially in the United States and
Soviet Union. After 1947, the potential of supersonic military
flight became the new yardstick for tactical and strategic military combat aircraft.

The first fighter capable of level flight past the Mach began
development in 1949 from the fertile imaginations of Ray
Rice and Ed Schmued of North American. Knowing this was
the next threshold to cross, and keenly aware of parallel Soviet development, they began work on the NA-180, also called
"Sabre 45." In addition to supersonic flight, the new fighter
would have an afterburning Pratt & Whitney J57 jet engine and
45-degree swept wings. This engine necessitated a longer, thinner, and heavier fuselage and, though the wingspan only increased by twenty inches, the overall wing area was 100 square
feet *larger* than its smaller brother. A fascinating example of

concurrent advancements, the design also featured the first use of titanium with milled, internal wing stiffeners instead of spars. This made the fighter more survivable from battle damage and also very, very strong while maintaining a relatively light weight. First of the "Century Series" fighters, it was redesignated by the USAF as the F-100, and NAA called it the "Super Sabre" to honor its lineage.*

Of course, the Soviet Union could not stand idly by while the West, specifically the United States, had such a capability and they did not. Utilizing captured data, especially the Focke-Wulf Ta 183, the Lavochkin design bureau flew its La-176 in the fall of 1948. A blunt-faced, swept-winged, and swept-tailed prototype, it rather resembled a stubby F-86, though the Russians mounted the wings high on the fuselage. By late December the Soviets claimed that test pilot Ivan Fedorov had flown the jet past Mach 1. More designs followed: the MiG-15/17; the La-15; and the YAK-23. By 1951 the Soviets were developing their own supersonic fighter through Mikoyan-Gurevich called the I-340: the MiG-19.

So it began.

On November 20, 1953, NACA test pilot Scott Crossfield reached Mach 2 in a Douglas D-558-2 Skyrocket: 1,291 miles per hour. Yeager set out to immediately beat him in the X-1A, which he briefly did by reaching Mach 2.44 on December 12, 1953, but Chuck nearly died himself due to a unique high-speed phenomena called *inertia coupling*. This can occur when the forward momentum of a high-speed fuselage exceeds the

---

* Century Series jets all have designation derived from "100": F-100, F-101, F-102, etc.

capability of the wings and/or horizontal tail surfaces to control the aircraft. Basically, the control surfaces are too small, and uncontrolled pitching, rolling, or yawing results, which create forces far beyond the structural limits of the aircraft.

Three years later USAF Captain Milburn "Mel" Apt broke Mach 3, nearly 2,100 miles per hour, in a Bell X-2. Due to inertial coupling he lost control of the aircraft on the way back to the dry lake and was tragically killed. A high-stakes technological leapfrog then ensued that, by necessity, had to continually develop better aircraft that flew faster and higher than whatever was being designed to shoot them down. This begat air-to-air missiles, air-to-surface missiles, and surface-to-air missiles, which then necessitated more advanced systems to detect and target fast-moving, supersonic aircraft. It was, as fighter pilots say, "a self-licking ice cream cone." Radar technology, used since World War II, was also rapidly evolving and soon a fighter would be at a tremendous disadvantage without its own onboard air-to-air radar. Fueled with the knowledge that virtually anything was now possible, design and capability began a dance that continues today.

If we can dream it, we can build it, became a credo still very much in effect. Surely there is no greater expression of this attitude during the decades after 1947 than the tantalizing possibility of space travel. And why not? If the demon could be chased to the far reaches of our atmosphere, then he could be pursued even farther into the vacuum of space. On October 4, 1957, a polished little globe, only twenty-three inches in diameter and festooned with antennas, was launched into low earth orbit from Site Number 1 in Soviet Kazakhstan. Called Sputnik 1, it completed 1,440 orbits of the earth until burning up twenty-one days after its launch. Though lagging somewhat in terrestrial technology, the Russians caught the West by surprise and triggered yet another dimension of the military-political-ideological arms race. The United States countered on January 31, 1958, as Explorer 1 blasted into orbit from Cape Canaveral, Florida, and another period of demon chasing began in earnest.

That same year saw President Dwight Eisenhower sign the National Aeronautics and Space Act, thus replacing the old NACA with a new organization centered on the National Aeronautics and Space Administration, or NASA. Among other goals, the new entity was charged with:

1.   The expansion of human knowledge of phenomena in the atmosphere and space.
2.   The improvement of the usefulness, performance, speed, safety, and efficiency of aeronautical and space vehicles.
3.   The preservation of the role of the United States as a leader in aeronautical and space science and technology . . .

In anticipation of a space race, the USAF initiated the MISS Program (Man in Space Soonest) in 1958, though it was canceled when NASA announced Project Mercury later that year. The goal was to put a man into orbit, and at the direction of President Eisenhower, the first American astronauts would be military test pilots. Out of 508 applicants, 7 were eventually chosen: Major John Glenn from the United States Marine Corps; Lieutenant Scott Carpenter with Lieutenant Commanders Wally Schirra and Alan Shepard from the Navy; and Major Deke Slayton along with Captains Gordon Cooper and Gus

Grissom from the U.S. Air Force. In addition to multiple psychological and physical screening programs, each man had to be under forty years of age, no taller than five foot eleven (due to the capsule size), and have at least 1,500 hours of flying time with a jet qualification. Additionally, each man had to have at least a bachelor's degree, or a professional equivalent: Al Boyd's faith in military test pilots was once again well deserved.

Less than three years later NASA's goal of American preeminence was directly challenged by the Soviet Union, who put the first man into space on April 12, 1961. Also launched from Kazakhstan's Baikonur Cosmodrome, Yuri Gagarin orbited the earth in Vostok 1 during his one hour and forty-eight-minute flight.* Less than a month later forty-five million Americans watched as astronaut Al Shepard blasted off on Mercury-Redstone 3, better known as Freedom 7, and the United States was truly in the space game. That same year saw USAF Captain Bob White pilot the North American X-15 to an astonishing Mach 4 on March 7; Mach 5 on June 23; and Mach 6, over 4,000 miles per hour, on November 9, 1961.

White had flown Mustangs with the 355th Fighter Group in World War II until shot down on his fifty-second mission during February 1945. After the war he finished a bachelor's degree in electrical engineering from New York University, then went on to George Washington University for his MBA. Back in the Air Force for Korea, he flew fighters out of Japan, and after graduating from the USAF Experimental Test Pilot School, he became the primary pilot for the X-15 program. On

* Fédération Aéronautique Internationale: Aviation and Space World Records.

July 17, 1962, Major White took the X-15 and chased the demon 59.6 miles above the earth to an altitude of 314,750 feet. The possibilities, it seemed, were endless.*

·Supersonic flight was quickly eclipsed by faster supersonic flight, then the wonder of the space program and, always, more war. Through hallway discussions and cocktails in Manhattan, members of the UN succeeded in preventing nuclear holocaust, yet the Cold War umbrella covered smaller eruptions in Korea, Vietnam, China, and the Middle East, to mention just a few. These were hardly on the scale of the last world war, but deadly enough to those fighting, and they always possessed the potential to spread. Through it all the pilots and engineers went on with life, some remaining with various programs, others retiring to new lives and, as always, some perishing while chasing the demon.

Having passed up the X-1 program to go to Princeton, Glen Edwards arrived back to Wright Field (now Wright-Patterson Air Force Base) with a master's of science in aeronautics and immediately jumped back into testing. He had been the primary project pilot for the Convair YB-46, and by May 1948 was out at Muroc with the Northrop YB-49: the Flying Wing. An amazing aircraft designed to fly nearly 10,000 miles and deliver nuclear bombs, it could have revolutionized long-range strategic warfare. But during performance testing on

* Bob White would go on to fight in his third war with his old unit, now the 355th Tactical Fighter Wing, and flew another seventy combat missions out of Takhli, Thailand, during Vietnam.

June 5, 1948, the outer wings collapsed and detached, killing everyone aboard, including Glen Edwards: he was thirty-two years old. * In his honor, on December 5, 1949, Muroc Air Force Base was renamed Edwards Air Force Base and remains so today. Glen is buried in Lincoln Cemetery, Lincoln, California (Plot 140, Grave 1).

Colonel Al Boyd would remain in Dayton until his promotion to brigadier general during September 1949. Given command of the Air Force Flight Test Center at Muroc Air Force Base, he immediately began preparation to move most of the military test operations to California. The transfer became official on February 4, 1951, and operations commenced from an old hangar on the South Base. Boyd also successfully lobbied to have the base renamed in honor of Glen Edwards, whom he regarded as an excellent test pilot and example to those who would follow. Shortly thereafter, the Air Force created the Air Research and Development Command (ARDC), which would oversee all research and development projects, as well as the new Experimental Test Pilot School.

Under Boyd, the name would again change to the U.S. Air Force Experimental Flight Test Pilot School, with a more stringent selection process. After adding another star to his shoulders, Al Boyd came back to the Wright Air Development Center in 1952, then went on to Air Research and Development Command headquarters. He retired in 1957, after flying 23,000 hours in 723 different types of aircraft. Boyd was vice president

---

* XB-49 #42-10238.

of the Westinghouse Defense & Space Group, then moved on
to General Dynamics in Fort Worth, and finally became a con-
sultant for Avco Lycoming. In 1963, the general flew solo from
Wichita, Kansas to Geneva, Switzerland. Al Boyd died twenty-
nine years to the day after the creation of the United States Air
Force: September 18, 1976. He is buried at Arlington National
Cemetery (Section 11, #733-1).

Chalmers "Slick" Goodlin, who could have flown supersonic anytime in 1946 if so directed, left Bell and the
United States to fight for Israel during the 1948-1949 Arab-
Israeli War. As a *mahal*, a foreign volunteer, Goodlin survived
forty combat missions in the LF Mk IXe Spitfire, and then he
flew in thousands of Jewish refugees for Near East Transport
before becoming a test pilot for the Israeli Air Force. A highly
successful businessman, Chal was a founder of Transavia, a
Dutch charter company that later emerged as a KLM subsidi-
ary. He also owned Seychelles-Kilimanjaro Air Transport and
was later the CEO of Burnelli Aircraft. Slick Goodlin died of
cancer in West Palm Beach, Florida, on October 20, 2005, and
is buried at St. Johns Cemetery, Hempfield Township, Pennsylvania.

Chuck Yeager remained in the Air Force. In 1949 he made
an attempt to take the X-1 off from the ground and fly past
the Mach but it failed, just as predicted, due to lack of fuel. He
had already been awarded the Collier and Mackay Trophies and
would later be presented with the Harmon Trophy. Yeager flew
chase for Jackie Cochrane when she became the first female

pilot to fly supersonic, and he would become the first commandant of the USAF Aerospace Research Pilot School in 1962.
In 1975 he retired from the USAF as a brigadier general after
thirty-three years on active duty; Yeager has since consulted for
video-game flight simulators and been the televised face of AC
Delco batteries. With the help of Leo Janos, he had an autobi-
ography published in 1985 called *Yeager*; as of February 2018,
Chuck Yeager is ninety-five years old.

It has been said that only the good die young, and this very
often applies to pilots like Jack Ridley. Unquestionably the
brains behind the USAF subsonic flight test team, Ridley
stayed with the X-1 until 1948, then transferred to the XB-47
program. This swept-wing, supersonic jet bomber was the
cornerstone of the new Strategic Air Command's response to
a Soviet nuclear threat, and Jack's practical, problem-solving
genius was irreplaceable in getting the aircraft to operational
status.

He came back to Edwards AFB after this and remained
in place until 1956. During this time, Ridley worked the X-2
through X-5 experimental programs, and the B-52 advanced
bomber. As chief of the Flight Test Engineering Laboratory, the
procedures he created while fleshing out these aircraft are still
in use today. "Jack Ridley was a good pilot and brilliant engineer," his wife, Nell, would later write. "But he was somewhat
forgetful about some the ordinary, everyday things in life. One
day he went to work with no insignia on his uniform." Yet his
mind never really stopped working. At the Officers' Club one
night, while engrossed in a conversation with one of his pilots,
he scribbled a few figures on a piece of paper and handed it

to the next table. Four MIT grad students had been agonizing over a problem and, while barely listening, Ridley gave them the solution.

In 1956, as a testament to his flying and engineering skill, Theodore von Kármán himself nominated the forty-year-old lieutenant colonel to the Flight Test Techniques Panel, part of the Aeronautical Research and Development team charged with consolidating efforts from all NATO countries. A year later, Jack was promoted to full colonel and sent to Japan as part of the U.S. Advisory Group. On March 12, 1957, while flying copilot on a C-47 inbound to Tokyo in bad weather, he died when the aircraft smashed into Mount Shirouma. Colonel Jackie Lynwood Ridley was posthumously inducted into the National Aviation Hall of Fame, and the Aviation Walk of Honor. He is buried in Arlington National Cemetery (Section 1, 236-D).

The boom rolled over the airport, broke windows, and sent the gathered reporters scrambling for cover after the silver fighter streaked past. At less than 1,000 feet, in full afterburner and already past Mach 1, George Welch thundered over the Palmdale, California, airport on May 25, 1953. Six years and seven months after taking up the first F-86 Sabre, he was showcasing North American's newest jet: the YF-100A Super Sabre. Now the company's chief engineering test pilot, George had spent some months in Korea during the war as a technical adviser for the F-86. Rumor has it that he flew over twenty combat missions as an "observer," something not so difficult to arrange for a World War II ace a few years after that war ended. It is also said that Welch shot down a few MiG-15s, though he was as indifferent to keeping score in Korea as he had been in the Pacific so, if true, others got the credit.

The new Super Sabre had problems, though. The military had initiated a program that would speed up the acquisition and fielding of new weapons and, like most bureaucratic solutions, looked good on paper but was short on practical reality. For aircraft procurement, the Cook-Craigie policy essentially meant a new design would be "flown" off the drawing board, and subsequent flight tests would prove the design. This was a major departure from validating the aircraft through a prototype before building production models. Military test pilots like Pete Everest and Chuck Yeager both concluded that the YF-100A had a slow engine response time at low speed and, more dangerous still, longitudinal stability problems. The rudder and horizontal stablizers were too small to deal with transonic airflow burbles, which could induce inertia coupling.

George disregarded the well-found warnings and on Columbus Day, October 12, 1954, he took the ninth production F-100 (#52-5764) on a high-speed, high-g test mission. One item on his test card was a symetrical, 7-g pull at 1.55 Mach, and when he did this, the smooth supersonic flow on top of the wing came apart, resulting in a burbling, disturbed mass of air that blanked out the vertical tail. In an instant the jet began yawing with no way for George to control it, and the fighter rapidly went out of control. Using over 300 pounds of force on the rudder (according to the flight data recorder) he tried to keep the F-100 flying straight but the oscillations deepened. At supersonic speeds this generated forces beyond the aircraft's structural limits, and the nose broke off.

Somewhere during these few seconds George reflexively pulled the handle and ejected, but not before the canopy bow and instrument panel smashed into his chest. Still traveling at over 700 miles per hour, George came out of the shattered fighter around 20,000 feet and plummeted straight down according to a bomber transiting the area. The parachute deployed, but the force ripped out several panels and he hit the ground hard. Fellow test pilots J. O. Roberts and Bob Baker flew to the crash site and found Welch still alive: barely. He died there on the desert a few minutes later.

George Welch remains an enigma to this day.

A superb fighter pilot, George fought his part of World War II with outclassed aircraft against the best of Japan's aviators, and he prevailed. He came home sick, exhausted, and victorious when others were just getting started. As a test pilot, George was skilled and confident to the point of recklessness, which is not the path to survival in high performance jets; and he did not survive. For all of his short, adventurous life Welch walked a different path; a prankster and a maverick, he was unquestionably courageous and thoroughly independent.

Yet there was something else eating away at him.

George was one of those lost souls who never made it all the way back from war. Growing up in the Great Depression he had learned how quickly everything could change and, like all combat veterans, he knew the frailty of life, and seemed determined to enjoy as much of it as possible while it lasted. On the day he was killed, his Pearl Harbor wingman Ken Taylor, now an Air Force colonel, was coming to California to persuade Welch to pass the flying torch on to others and take a desk job. His time was up, and his friends were trying to save his life. It would not have worked and, in the event, was too late. George Welch is buried on a quiet hillside in Arlington National Cemetery, flanked by a pair of trees and only a few yards from the Tomb of the Unknown Soldier where, if there is grace for fighter pilots, he may finally be at peace.*

Ken Chilstrom finished the Sabre's Phase II testing and was the original American military F-86 pilot. The next year, 1948, was a busy time for Ken. He became the first USAF/ USN exchange pilot and spent several weeks at Naval Air Station Pensacola getting carrier qualified. In a satisfying twist of fate, he learned the Navy art of takeoffs and landings on the USS *Wright* (CVL-49), a Saipan-class light carrier named for the Wright brothers.† Ken eventually racked up fifty traps—carrier landings—flying F8F Bearcats with Carrier Air Group Seven aboard the USS *Leyte Gulf* (CV-32).

Returning to Ohio, he was appointed as commandant of the USAF Test Pilot School and enjoyed roaring around Dayton in a bright red convertible with white sidewalls. Following the success of the X-1 and XP-86 programs, the world of military test pilots was expanding rapidly. Busy as he was, Ken also managed to fall in love. Miss Ruth Bertsch worked for the base public relations office, so she had heard it all from the hotshot pilots at Wright-Patterson. Nevertheless, within a year he proposed to her at New York's Waldorf-Astoria and they were married. The Air Force then received a request from Warner Brothers to provide a technical adviser for an upcoming film about test pilots, and Ken was selected. He, Ruth, and the red convertible got to spend four months in Hollywood with Hum-

---

* Section 6, 8578-D.

† The first USS *Wright* (AV-1) was named for only Orville Wright.

phrey Bogart, Eleanor Parker, and Raymond Massey shooting *Chain Lightning*: it was released in 1950.

Ken was then picked to represent USAF test pilots with their Royal Air Force counterparts, and over the course of several months in the United Kingdom at Boscombe Down and Farnborough he flew twenty-five British aircraft. This year saw the birth of his first son, who he named Glen for his fallen friend Glen Edwards. Ken left for Japan in December 1951 to be the fighter requirements officer for the Far East Air Force (FEAF), returning to the Pentagon three years later as a lieutnant colonel with another son: John Scott Chilstrom. Spending the next four years managing Century Series fighter programs like the F-100 and F-105, Ken was promoted to full colonel and retired in 1964 after twenty-five years on active duty. That same year a daughter was born, Carol Lynn, and the family moved to Valley Forge, Pennsylvania.* For the next twenty-two years he held executive positions with General Electric, Boeing-Vertol, and others before again retiring, this time in 1986 from Pratt&Whitney.

Ken Chilstrom has been chasing the demon his entire life; through the Roaring Twenties, the Great Depression, the Second World War, and into the space age. He nipped at its heels through 147 different types of aircraft and thousands of hours flown during a marvelously exciting career. As his ninety-seventh birthday approaches in April 2018, Ken has not slowed down much and can look back on an equally successful, well-lived life. Every day he can see the effects of the ripples that

---

* Glen Chilstrom became a noted psychiatrist; John Scott Chilstrom followed his father into the USAF and retired as a colonel; Carol Lynn is a successful businesswoman.

he, and those like him, caused by their endurance through the 1930s, their victory in the war, and their courage at the onset of the greatest age aviation has ever known. He was a deadly warrior and skilled pilot who became a loving husband and a superb father; Ken Chilstrom was, and is, a good man.

His legacy, his code, is a simple one: faith, family, and love of his country. "I was honored," he said, "to serve with eagles." If the chase for the demon is also a pursuit for one's deepest desires, then Ken did indeed catch the demon, and in doing so won his own race.

Acknowledgments

Once again, I owe my profound thanks to Peter Hubbard, executive editor at William Morrow/HarperCollins, for his unquenchable belief in each of my books, and the immense amount of time, effort, exasperation, and perspiration he pours into each work. Without his expertise, rampant enthusiasm, and tolerant nature, none of these stories would be possible. I am grateful, though do not always say so loudly enough, to his associate editor, Nick Amphlett, and my publicist, Maria Silva, for their unstinting devotion to making these books a success. I can neither imagine nor wish for a more dedicated group of professionals than the entire HarperCollins/William Morrow team.

Special thanks to George Marrett, former test pilot and fellow author, for putting me in contact with the Chilstroms; Colonel John Scott Chilstrom, a brother pilot, was kind enough to arrange an introduction for me to his distinguished father, Colonel Ken Chilstrom, without whom this book would lack the rich detail and untold stories he provided.

My gratitude and admiration go out to the tireless professionals of our National Air and Space Museum. Namely, Dr. Alex Spencer, Curator of the Aeronautics Department, who took me to the museum's amazing Suitland, Maryland, storage facility and permitted me to crawl around in our national treasures; Dr. Bob van der Linden, author and chairman of the Aeronautics department, who cheerfully tolerated my pesky questions and allowed me sift through reams of priceless original documents, transcripts, and reports; finally, my sincere respects to Dr. John Anderson, curator of aerodynamics, for his patience, boundless knowledge of all things aeronautical, and willingness to sacrifice his invaluable time. Dr. Anderson can count eight critically acclaimed books among his lengthy accomplishments, as well as membership in *Who's Who in America,* a Glenn L. Martin Distinguished Professor for Education in Aerospace Engineering appointment, and Professor Emeritus status for the University of Maryland.

Dr. Mary Ruwell, archivist for the United States Air Force Academy, granted me access to her wonderful collection, and was tremendously helpful with the aviation history aspects of this book, as was Dr. John Terino of the USAF Air Command and Staff College, who is always willing to critique my writing and save me from public errors. Similarly, Mike Dugre of the USAF Historical Office, and Dr. Roger Launius of Launius Historical Services were instrumental in filling in the blanks for the NACA's early history. Michael Lombardi, Boeing's corporate historian, was kind enough to dig through his archives and send me the North American Aviation Test Reports filed by George Welch during the fall of 1947. Another fellow author, Lauren Kessler, now a professor at the University of Oregon, went beyond professional courtesy to personally answer my questions regarding Pancho Barnes.

Last, but never least, my parents; my wife, Beth; my children; and the rest of my family must be recognized for enduring the perils of living with an author. Uncooked dinners and unmade lunches, neglected projects, forgotten appointments, and dozens of other transgressions for which I alone am responsible, and for some reason they continue to forgive. In partial penance for this unsettled life—which none of you asked for—please accept my lasting apologies, love, and appreciation.

Glossary

AAF: Army Air Force (United States).

ACCELERATION: a change in velocity over a specified time.

AERODYNAMICS: the study of air and the forces produced by its behavior.

AILERONS: rectangular surfaces on a wing's trailing edge used to control rolling.

ALLIED POWERS: nations allied against the Axis during World War II; primarily the United States, the British Empire and her Commonwealth of Nations, China, and the Soviet Union.

ANGLE OF ATTACK: the longitudinal axis of an airfoil with relation to the relative wind.

AREA RULE: narrowing a fuselage at the wing juncture to reduce drag.

ASPECT RATIO: a ratio between the square of the wingspan and its total area.

ATTITUDE: the orientation of an aircraft in relation to the horizon, or any fixed reference.

AXIS POWERS: National Socialist Germany, Fascist Italy, and Imperial Japan. The name is taken from the Rome–Berlin–Tokyo Axis.

BIPLANE: an aircraft with two sets of fixed wings.

CAMBER: the curve of an airfoil surface.

CENTER OF GRAVITY: an imaginary reference point for the center of mass within an object

CHORD LINE: a notional line running through a wing connecting the leading and trailing edges.

COMPRESSIBILITY: the change in volume of a solid or fluid in response to pressure.

CRITICAL MACH NUMBER: the lowest speed that some point on an aircraft becomes supersonic.

DIHEDRAL: in aerodynamics, an upward angle between two surfaces.

DoD: Department of Defense.

DRAG: in fluid dynamics, the cumulative resistance of everything affecting forward movement. For an aircraft, thrust must exceed drag for forward flight.

EXPERTEN: a somewhat ambiguous term for a Luftwaffe top scorer. Roughly equivalent in prestige to an Allied ace.

G-FORCE: for a human, this is an acceleration against gravity that is felt as weight.

IAS: indicated air speed.

JET: an engine that generates forward thrust by expelling high-pressure exhaust created through the combustion of compressed air.

JPL: Jet Propulsion Laboratory.

LIFT: an upward force produced by pressure differentials acting on an airfoil.

LOAD FACTOR: the ratio of generated lift to the overall weight of an aircraft.

MACH NUMBER: an aircraft's velocity measured against the speed of sound.

MONOPLANE: an aircraft with single set of fixed wings.

NACA: National Advisory Committee for Aeronautics.

NASA: National Aeronautics and Space Administration.

PITCH: a nose-up or -down attitude of an aircraft.

RAMJET: an engine that functions with no compressor. Forcible combustion occurs during flight as air funnels through a narrow tube.

SHOCK WAVE: an expansion of energy that propagates outward and results in a sudden change of density, temperature, and pressure.

SPEED OF SOUND: the distance traveled by a sound wave over a unit of time; at 32° F this is 1,087 ft/s, or 741 mph.

STALL: a point on an airfoil where airflow separates and effective lift ceases to be produced.

SUPERSONIC: movement of an object, or air around an object, faster than the speed of sound.

TAS: true air speed.

TRANSONIC: a notoriously ambiguous region of severe stress between an airfoil's critical Mach number and approximately Mach 1.2 where airflow is erratic, unpredictable, and dangerous.

WEIGHT: the combined mass of an aircraft and everything in it. Must be overcome by lift for flight to occur.

WING AREA: wingspan multiplied by the chord.

WING LOADING: total aircraft mass divided by the wing area. Essential in calculating lift and the overall maneuvering performance of an aircraft. The faster an aircraft flies, the more lift is produced by the wing, so a smaller wing can carry the same weight in level flight. However, greater speeds must then be maintained to produce lift.

WINGSPAN: distance from wingtip to wingtip.

YAW: A rotation around a perpendicular axis; for an aircraft, this is a side-to-side movement of the axis between the nose and tail.

Appendix:
Aerodynamics 101

Before venturing deeply into the history and mysteries of aviation, it would be advisable to review some basis concepts that, while fairly commonplace in today's world, are fundamental to flight and were misunderstood for centuries. To begin with, an *airfoil* is the shape, seen in cross section, of a body that moves through a fluid. For our purpose this usually means a wing of some type, and the fluid is air, though the terms can and do apply elsewhere.

Nevertheless, it is this movement that produces an aerodynamic force as the fluid splits around the airfoil. We tend to consider fluid as being liquid, but it can be anything really, as long as its component molecules flow freely, such as air; or it can be something that assumes the shape of its container, like a gas or liquid. For our discussion we will use air, though visualizing water often aids in aerodynamic explanations as water can seen and air usually cannot. Aerodynamics, then, is a study of how bodies move through air; and air, according to NASA, is "a physical substance which has weight. It has molecules which are constantly moving. Air *pressure* is created by the molecules moving around. Air is a mixture of different gases; oxygen, carbon dioxide and nitrogen. All things that fly need air."

So an airfoil can be your hand protruding from a car window, or a blade, feather, a wing—whatever is moving through the fluid. We will usually be discussing a wing, which is a specific airfoil, though keep in mind that propellers and turbines are also airfoils. For a symmetrical airfoil—one shaped equally on the top and bottom—the air would move across both surfaces at the same speed. The curve, or *camber*, of an asymmetrical airfoil is not the same on the top and bottom, and because mass (in this case airflow) must be conserved it must therefore be the same at any point along the airfoil's cross section. In order to do this, the flow velocity over the top, cambered section is faster than that below. This results in less pressure being exerted downward as the higher-velocity molecules are dispersed, creating "thinner" air and less pressure. Slower air is "thicker" and produces greater pressure under the wing, which pushes upward and, meeting less resistance, correspondingly *lifts* the airfoil.

This relationship between velocity and pressure was well understood for centuries, at least as far as wind and water were concerned. Mariners applied it to the setting, or trim, of a sail (is really just a movable airfoil) for varying the pressure to move the ship that was attached to the sail. In 1738, Swiss mathematician Daniel Bernoulli articulated the principle in his *Hydrodynamica*, though its adaptation to the fledgling world of aviation would take a bit longer. Still, it is a simple notion. A strong pressure pushing upwards against a weaker pressure will continue to move in that direction and take the airfoil, or wing, with it.

Camber can be, and almost always is, designed into an air-foil to intentionally produce the pressure differential that creates lift. This curve, and its effects, can also be altered by the use of leading-edge and trailing-edge flaps, which give the wing different lifting qualities during specific flight areas, and by ailerons, which we discuss in detail later.

John Smeaton, an English civil engineer, discovered in 1759 that lift was greater when an airfoil was cambered rather than flat. This seems to have been largely forgotten and though the concept was grasped during the nineteenth century, it was not patented until 1884 by Horatio Phillips, who called it a "double surface airfoil."

"The particles of air struck by the convex upper surface," reads the patent text, "are deflected upward, thereby causing a partial vacuum over the greater portion of the upper surface. In this way a greater pressure than the atmospheric pressure is produced on the under surface of the blade." This was a significant revelation, yet one that already intuitively occurred to several aviation pioneers. In any event, the properties were quickly grasped, improved, and incorporated into relatively efficient wings capable of producing *lift*, and it is this lift that enables flight. It does not, by itself, constitute flying, but without this force the other components of flight—weight, drag, thrust, and control—are generally impotent, at least as far as flight is concerned. However, the generation of lift alone does not mean the bird, insect, or aircraft will get off the ground; for that to happen the lift must be greater than the *weight* of whatever is being lifted. Weight is a gravitational force, a vector, so it must have both magnitude and a direction. Magnitude is a function of the combined mass from the pieces, parts, people, and the aircraft itself. Its direction is some component angle opposite the lift vector as gravity tries to pull the mass into the center of the earth. Bottom line: a wing must produce enough lift to overcome the associated weight in order to physically get itself and its attachments off the ground.

This is still not flying, but we are getting closer. *Thrust* is simply any force that propels a craft through the air. A bird flaps its wings, getting the air moving over hundreds of tiny airfoils in the feathers, and thus producing lift. A glider wing can do this as long as it has forward velocity, but a glider has no way to produce thrust itself so man needed to artificially lift it. This began with the steam engine then continued with internal combustion, the rocket, and finally the jet.

Just as lift must be greater than the weight for a craft to become airborne, so must thrust exceed *drag* for the craft to move. Drag is the cumulative resistance of everything affecting the forward movement of a craft through the air. Primarily this is due to variations in air pressure and friction, acting locally all over a body. All the pieces, parts, and people on, or inside, the aircraft: everything generates drag of one type or another. For example, contact between air molecules and the surface of an aircraft causes a shear-stress, or friction, and this **skin-friction** is a major source of drag. Rubbing your hands together very fast illustrates the point; you feel heat, which is a result of friction.

Shape is also critical; visualize a torpedo moving through the water versus a block of wood, and the concept of **form** drag is clear. Pressure acting on a body and the velocity of the flow around it are not constant values; they are localized due to many factors, including the construction of the craft itself. Another type of drag is caused by **interference**: the physical joining of different parts of the aircraft, such as the fuselage meeting the wing. These have always been major design concerns, and re-finement continues even today. Collectively, the combination of skin friction and form drag is termed **parasite** drag, as it is not caused by lift, but by the construction and design of the aircraft itself.

On the other hand, **induced** drag is a by-product of lift. Most of this is derived from revolving areas of flow, like smoke rings or miniature tornadoes, which are called vortices. These are created by the pressure differential that generates lift, and they spill over the wingtips. Now, the divergent airflow that permits lift also merges at some point downstream past the wing where the flow rejoins. The impact of both upper and lower flows crashing into each other produce vortices, which in turn create a disturbance and more drag. So the parasite and induced figures added together become **total drag.** Other types of drag, wave and compressibility drag specifically, occur as airflow becomes transonic and supersonic, but this will be discussed later.

However, the pictured airfoil is idealized. It is moving straight ahead, parallel to the horizon and directly into the rel-ative wind. Real wings, especially those on highly maneuver-able fighters, are very rarely level and all aircraft will vary the angle at which they strike the relative wind as they take off and land. This angular difference between the airflow and the wing is called the angle of incidence or, more commonly, the **angle of attack**. It is critical because this affects all the aerodynamic forces described above, and especially the amount of lift that is produced when the flow impacts the wing at angles other than zero. This calculation, vital to wing and aircraft performance, is expressed as a ratio of **Lift over Drag**, and reveals that lift increases as the angle of attack decreases. So the more directly a wing strikes air, then the greater pressure, and lift, will be generated because of it.

But what happens when the angle increases, as it does in slow-speed flight during landings, or during extreme maneuvering such as combat? Well, the wing continues to produce lift with varying degrees of efficiency until it reaches a **critical angle of attack**, the point where its maximum lift is produced. Visualize peeling an orange and feeling the skin finally ripping off in your fingers; this is what happens to the airflow when it finally separates from the top of the wing.

Beyond this point the wing stalls, just as a sail will flap ineffectively when pointed into the wind. When this occurs, the angle of attack must be decreased in order for air to move over the wing again and generate more lift than drag. This took some time to work out and did not fully mature, as with many other aerodynamic principles, until the advent of aerial combat during World War I.

Once aerodynamic forces were better understood, the next logical step was to account for each of these in aircraft development, and much of this effort toward refinement went into wing design. Benjamin Robins, an eighteenth-century English Quaker turned military engineer, noted that variously shaped airfoils produced different aerodynamic results. He was the first to develop the concept of a **wing aspect ratio** and relate it to wing design. This ratio is simply the wingspan squared divided by the planform area or, in the case of a rectangular wing, planform area. It revealed that in *subsonic* flight a rectangular wing, or one with a high aspect ratio, produces more lift than a stubby, low aspect wing, as there is more lifting surface available.

Of course, there are always aerodynamic trade-offs. High

aspect wings produce more lift at the expense of structural stability; they are longer and protrude farther from the fuselage so in the early strut-and-wire days, extensive bracing was required. Longer wings also shift the pressure outboard, away from the fuselage, and as speed increases this affects controllability. Shorter, low aspect wings are much stronger, but do not generate lift to the same degree. The initial compromise was the development of biwing aircraft—the biplane. A *pair* of shorter wings permitted greater lift and increased strength, while keeping the aspect ratio fairly high. It is well to remember that comprehending, and putting into practice, these aerodynamic truths took years of trial and error, with spectacular failures and eventual successes.

The monoplane quickly replaced the biplane once technology and design evolved where a single, structurally strong, high aspect wing could be used in place of two wings and this resulted in much less drag, as there were no exposed struts and wires. Corresponding engine advancements generated higher power, and the increase in speed produced greater lift so smaller wings could be effectively used. These weighed less and generated less drag, so the more powerful engines now produced excess thrust. Wing design is a perfect example of the advantages for parallel development, in that the larger wing also permitted multiple gun mounts, with more interior space for ammunition storage and retractable landing gear—all made possible by aerodynamics.

Another component needs to be defined and, though not a force per se, **control** is essential for true flight; that is, the capability of an onboard pilot to physically and deliberately manipulate an aircraft through multiple dimensions. There are three primary axes by which this occurs. **Pitch** is along the lateral axis running from wingtip to wingtip, and the vertical up or down action is initiated by a stick or yoke in the cockpit. When a pilot pulls back, rectangular surfaces on the tail called *elevators* deflect upward into the airstream. The flow strikes the elevators and the resulting pressure pushes the tail down, which raises the nose of the aircraft. If the pilot pushes forward the opposite occurs; the elevators deflect down, which raises the tail and the nose falls.

Yaw is like swinging around a vertical pole. It is a rotation around the perpendicular axis and is controlled by a *rudder* that functions exactly like one on a boat. Installed vertically on the tail, it is moved left or right by pedals on the cockpit floor. Again, deflecting the airflow pushes, or yaws, the aircraft. Due to changes in lift, an aircraft cannot maintain level flight in a rudder-only type of turn, and yawing alone is a sloppy way to change direction, like skidding around a corner in a car, versus turning the wheel. To counter the loss of lift a pilot will **roll** the aircraft, using *ailerons*. These are smaller rectangular surfaces located on the trailing edge of the wings. Moving a stick or yoke left or right works the ailerons in tandem; one raises and one lowers, which decreases and increases the lift on both wingtips, respectively, and rolls the aircraft. It is through the combined use of all these controls that nonlinear oblique movements, just like a bird makes, are possible. Variations on these basic controls are as numerous as variations among aircraft themselves, and the flight control system of commercial airliner supports much different requirements than that of a jet fighter, yet they all work along the same principles.

So with these basic aerodynamic forces in mind, how is an airplane put together to fly? The **fuselage** is the center of all this, the main body and principal structural component of an aircraft. From the French *fuselé*, or spindle, it was originally a girder and lattice structure filled with bracing and wires. Wood was the preferred material, since it could be readily cut, shaped, and molded, and the entire arrangement was covered with fabric, usually linen or cotton. Various coatings were employed for waterproofing, as the sky is often a wet place. Unfortunately for many early combat aviators, wood is easily shattered by force, or bullets, and coated with varnished, painted cotton it readily burned.

As engines became more powerful not only were more secure mountings needed, but also a framework that could withstand the additional stresses resulting from increased performance. The original solution was a *monocoque*, or single shell, fuselage that carries all the aerodynamic stress. A mold for each half of the fuselage was created, and thin strips of wood were layered at right angles and glued to one another. Once dry, both halves were glued together and the resulting structure was very light and extremely strong. This was time consuming and difficult to repair, especially battle damage, so as with most aviation issues a compromise was reached.

The solution was a synthesis of both two methods; **semi-monocoque** or veneer, had a frame, spars, and cross bracing but was covered by wood panels rather than fabric This type of fuselage would survive the strut-and-wire biplane era, progress into the all-metal aircraft age, and various derivatives are still used today. Aft of the fuselage is the tail assembly, or **empennage**. This vertical tail section contains a rudder and the horizontal tail with elevators. Structures vary, depending on the type of the aircraft and its purpose, but the function is the same. Construction of the empennage plays a crucial role in the plane's balance, center of gravity, and aerodynamic per-

formance. This was particularly true in the early days of supersonic flight test as the disturbed airflow trailing from the
wings severely affected downstream airflow over the elevators,
which resulted in deadly control issues.

Despite knowledge of how these forces, axes, and controls
function together, they are, without a pilot, just so much wood,
metal, and cable. The pilot is the brain, the living computer,
who makes these diverse components work harmoniously to
produce flight, and it all comes together in the cockpit. The
sixteenth-century term denoted a pit where bloody fights took
place between animals or men and was later applied by mariners to describe a partially enclosed, sunken area on a ship's
deck occupied by the helmsman. As sailing and flying are both
practical applications of fluid dynamics, it is natural that they
share many terms: rudder, for example, or the concepts of pitch,
roll, and yaw. As aircraft fuselages became more enclosed, the
term cockpit seemed a natural fit for a small, semiprotected
space where the craft could be controlled by a supine pilot.

There was resistance to this.

The Wright brothers, and others before them, believed that
a pilot's proper position was prone. This intuitively made sense
to them for several reasons. First, a bird generally flies parallel
to the relative wind, which exposes less of its body to the airflow, decreases drag, increases speed, and permits the greatest
amount of air to flow over its wings. A prone position also allowed the Wrights to control the Flyer by shifting their weight
like a bird or bat, since their wing-warping method of control
was attached to the pilot's hips.

Just to the left of the engine the cradle was attached by cables
that ran outward to the wingtips, so by shifting his weight side-
ways the pilot would twist, or warp, the wingtips to turn the

plane. The Wrights utilized a pair of levers (visible forward of the cradle) to move the rudder and elevators, which were at the front of their Flyer, rather than on the tail. It quickly became apparent to other designers that a seated position was more natural for a pilot, and that a yoke or joystick was easier to manipulate than levers. By the end of the first decade of manned flight this had become a more or less normal configuration that was permanently solidified by the Great War.

Air combat is fast; it requires instant maneuvering in multiple dimensions under increasing gravitational, or "g," forces in order to aim and shoot. This new reality necessitated an efficient and relatively simple means of control since pilots had to be able to fight with their aircraft, not just fly it from point to point, along with navigating and managing weapons. As some 85 percent of humans are right-handed, it became understood that the stick, as the most sensitive control, would be manipulated by a pilot's right hand.* It was only natural then that the throttle control be placed on the left side of the cockpit for the pilot's other hand. This is a basic configuration that is still widely used today and adapted to all sorts of commercial, private, or military aircraft.

Great War fighters were highly visual machines as there were no aircraft-to-aircraft radio communications, and airborne radar was decades into the future. The Sopwith Pup is typical; a control stick with a firing button is operated by the pilot's right hand. The throttle quadrant is visible on the left bulkhead

---

* Figures vary and there is no single consistent percentage, though *Scientific American* reckons the number may be as high as 95 percent. It is agreed that the majority of humans are right hand dominant.

and rudder pedals are just visible beneath the rudimentary in-strument panel. By this time, especially in single-seat fighters, the importance of a sensible cockpit configuration was realized and being incorporated into new designs.

By the late 1940s and early 1950s, jet fighters like the North American F-86 ruled the sky; however, event though instru-mentation and weapons capability had greatly advanced, the ba-sic cockpit arrangement of stick, throttle, and rudder remained the same. The challenge of displaying and utilizing evolving technology to its best advantage is an ongoing process of refine-ment that also continues today.

Indeed, one of the most enduring aspects of aircraft develop-ment is that so much of it has remained consistent through the decades; enclosed cockpits and jet engines replaced older pis-ton models and open cockpit fighters, but these are technical advances resulting from man's innovativeness and external cir-cumstances like the world wars. Great leaps are rare and widely spaced, but when they do occur and a wall is breached or the demon is sighted again, such leaps are often as drastic as they are enduring. There are constants, however; aerodynamic prin-ciples do not change, though our knowledge of them increases and we may utilize them better; and, fortunately, men do not change. There will always be those who are more than willing to take what we know and scream off into the blue in search of what we do not know.

Notes

Chapter 1: Flying Monks to Mud Ducks

Dr. Bob Van der Linden and Dr. Alex Spencer of the National Air & Space Museum were both very helpful with procuring the relatively obscure information needed for this chapter. There is a Ba-349 Natter file that I was permitted to sift through, including the original NATTER INTERCEPTOR REPORT (July 1945) compiled by Dr. Clark Millikan for the Combined Intelligence Objectives Subcommittee. This mainly focuses on tactical applications and armament, though there is also an excellent photograph of the one Natter procured by the United States—incidentally, the same aircraft I was able to examine in Washington at the Air & Space Museum's Paul Garber Facility storage facility.

"Erich Bachem's Snake in the Sky" by Dr. Alfred Price (Air International, July 1996) is an excellent expansion of technical sources and organization of Natter field units. Price's article also details the launch, flight, and pilot separation sequence.

The Ministère de la Marine report (July 1939) from France's Technical and Industrial Second Bureau is an interesting intelligence evaluation of Germany's pre-war jet engine development.

Igor Witkowski's "The Truth About the Wunderwaffe" has a short section concerning the Natter (pp. 177–178).

The original writings of aviation pioneers are, of course, primary sources yet they are relatively inaccessible and sometimes challenging to read. These include Leonardo da Vinci's *On Floating Bodies* and *Codex on the Flight of Birds* where lift is truly discussed for the first time. Early treatises from Archimedes reveal his thoughts on pressure, which will continue to plague aerodynamicists through the early jet age. Excellent secondary sources, which encompass many ancient or renaissance works, include Dr. John Anderson's *A History of Aerodynamics* and *The Airplane: A History of Its Technology*. For those interested in the little known yet crucial early development of aviation, see Chapter 3, "Starts and Stops," of Anderson's *The Airplane*.

The U.S. Air Force Academy McDermott Library also has a comparatively small but eclectic assortment of aviation artifacts and writings. In particular, the Gimbel collection, a 1970 bequest from the famous department store family, is well worth perusing. Dr. Mary Elizabeth Ruwell, Chief of Special Collections, is extremely knowledgeable, amiable, and she was willing to let me sift through the material while answering my pesky questions.

17   **Though usually resulting:** See Anderson's *A History of Aerodynamics*, pp. 17–21. This contains an interesting summary of Archimedes' work on fluid statics that will, centuries later, significantly impact the study of pressure.

17   **In the early eleventh century:** *Test Pilots* (Hallion), pp. 2–11.

18   **On a moonlit summer night:** *Lords of the Sky* (Hampton), pp. 9–11.

19   **George Cayley, a self-educated baronet:** Richard Dee's *The Man Who Discovered Flight: George Cayley and the First Airplane* is an excellent source of information as is J. Lawrence Pritchard's 1954 summary in *Flight* magazine, also Anderson's *The Airplane*, pp. 28–32.

22   **Dr. Anderson says of Henson's monstrosity:** Anderson, *The Airplane*, p. 37.

27   **He ingeniously adapted the Pratt:** Octave Chanute's realization that "spreading the load" was essential to manned, human flight had a profound impact on the science of aerodynamics as this was a deliberate, thoughtful progression from the emulation of birds or bats. Humans are built differently, are much heavier, and use vastly different types of muscles, therefore copying flying creatures could only take our science of flight so far.

27   **On May 9, 1896:** Langley AFB obviously has prodigious archives related to decades of fascinating work there. Mr. Michael Dugre and Ms. Gayle Langevin were helpful with much of Samuel Langley's background information. See also Richard Hallion's highly informative *Test Pilots: The Frontiersmen of Flight*, pp. 23–26.

29   **a handful of mortar:** *Washington Post*, October 8, 1903.

30   **As with their predecessors:** Written histories of the Wrights is as voluminous as it is varied. Much of the older information is distinctly biased and paints the brothers in an unrealistically favorable light. I am not detracting from their achievements, but their failures were just as significant and rarely mentioned until more modern times. These shortcomings and their impact are discussed in some detail in Chapter Two.

31   **with great distinction:** *Aeronautical Annual,* "Wheeling and Flying."
1896.

Chapter 2: The Cauldron

37   **Ken Chilstrom was born during:** Personal interview at Ken's
home in Virginia. This was the first in a series of dozens of conver-
sations where he related his fascinating firsthand accounts of the last
nine decades.

37   **By the time Ken's father:** The happier aspects of the first few
years following the Great War can be found in a number of sources.
Among my top choices are Allen's *Only Yesterday,* David Kyvig's *Daily
Life in the United States,* and specifically the first five chapters of *Last
Call,* by Daniel Okrent. See also a previous work of mine, *The Flight,*
Chapter Two. for those truly wishing to delve deep, Margaret Macmillan's superb *Paris 1919: Six Months That Changed the World* takes the
reader through the unsettled time following the war, and the far-reach-
ing consequences from the Treaty of Versailles.

39   **The decade got off to:** Aviation was an exciting, integral aspect of
life in the 1920s, and I detailed several early transatlantic crossings in
The Flight, pp. 53–55. Part One of Philipp Blom's *Fracture,* specifically
"The End of Hope" and "Men Behaving Badly" are well worth reading
for insight into the emerging social fabric of the United States.

40   **This was the first nonstop:** *The Flight,* pp. 121–122.

41   **Most early engines were:** *Lords of the Sky,* pp. 57–59.

44   **All early forms of control:** A pivotal advance in one regime that
opens up a series of metaphorical doors in adjacent areas is a pattern
in aviation development. In this case, the aileron made it possible to
control faster and more maneuverable aircraft, which, in turn, was a
catalyst for powerful engines. These together made it possible for aircraft to be used as weapons, thus necessitating advances in structure,
design, weapons and tactics, and so on.

46   **For this reason the French:** Alberto Santos-Dumont is another
figure often overlooked, at least in the United States, for his contributions to early flight. Paul Hoffman's *Wings of Madness,* and the
monograph *Santos-Dumont and the Invention of the Airplane* are both
excellent sources.

47   **Some designers like Glenn Curtiss:** Curtiss is well worth researching for those intrigued by the early days of aviation, especially in
the United States. An astute businessman, Curtiss was well aware of
the value of publicity and promoted aviation till his dying day. Setting
an unofficial speed record of 136.36 mph on a custom V-8 motorcycle,
he was the world's fastest man from 1906 to 1911. Curtiss designed

and built the NC (Navy Curtiss) class of flying boats that first crossed the Atlantic in 1919. He was also responsible for a number of aviation firsts—retractable landing gear and aircraft pontoons—and was the original licensed U.S. manufacturer of aircraft. See Studer's *Sky Storming Yankee* and Roseberry's *Glenn Curtiss: Pioneer of Flight* for more information.

50   **Post could also see:** Augustus Post is well worth further research for aviation enthusiasts. A unique and eccentric man, there are numerous articles, court records, and writings concerning Post, but a good central source is his own papers in the Benson Ford Research Center of Dearborn, Michigan. Post also published poetry, authored a book called *Skycraft*, and composed an opera.

52   **A perfect storm of:** Without the turbulence of the 1920s it would certainly have been more difficult, but certainly not impossible, for fascism, national socialism, and imperialism to take hold as they did in Italy, Germany, and Japan, respectively. Though the economic collapse in the United States played a significant role in triggering the subsequent global recession, it was not a standalone event. The dominoes had been set up to fall through a tragic, albeit interesting set of miscalculations, incompetence, and rank ambition. See Allen's *Only Yesterday*, Chapters XII, XIII, and XIV; Martin Pugh's *We Danced All Night*, pp. 1–20; and *The Long Week-End* by Graves and Hodge.

54   **After a brief postwar:** Read Blom's *Fracture*, pp. 97–215, for a year-by-year synopsis of the early to mid-1920s. Though the language is a bit stilted, *Only Yesterday* remains a highly readable account of the same period, specifically pp. 137–195.

54   **By 1925 the radio:** The *Statistical Abstracts of the United States*, published by the U.S. Census Bureau, remain a wealth of painstakingly compiled hard information. Easily accessible through the Bureau's website, there are data and summaries for virtually every aspect of American life.

55   **But there were problems:** The onset of the collapse of 1929 had surprising roots and, like an earthquake, there were tremors long before the eruption. T. H. Watkins, in *The Great Depression*, outlines much of this through Chapter 1, "The Prologue Years," and Bill Bryson similarly works his entertaining accounts in *One Summer: America, 1927*, particularly pp. 357–429.

Chapter 3: The Next Leap
63   **By the mid-1930s:** Money, credit, buying power—these are the oils of all developed societies just as resources are the fuels. To continue the engine analogy, if one runs low, the forward momentum

stops or the engine comes apart in pieces. Lack of resources would be the often overlooked crisis facing the Axis powers and was a catalyst for their initial conquests during the first two years of the world war. For the United States, a decade's worth of overproduction and install-ment (credit) buying had grossly inflated American industry, so when combined with overpriced commodities and uncertainty in foreign markets, a reckoning was inevitable—but not necessarily cataclysmic. However, the failure of a commercial republic like the United States to prevent its banking collapse, and then to immediately correct it, is as astonishing as it is frightening. The cultural and societal damage shaped several generations, especially the men who chased the demon, and created the drastic consequences of the late 1930s that led to World War II. I suggest Watkins's *The Great Depression*; Chapters 7, 8, and 9, for those intrigued by the subject. Also Richard Overy's *The Twi-light Years* contains highly informative insights into Britain's condition during the same time, which is vital in understanding Whittle's diffi-culties producing the world's first jet engine.

65   **It began in the:** Kyvig's *Daily Life in the United States*, pp. 163–230, expands on the impact of the Great Depression in North Ameri-can society, and the results of government solutions.

65   **It was into this world:** Ken Chilstrom's account was fascinating. Of course, he learned later in life how it all started but his firsthand, personal memories of this crucial time and its day-to-day effect on his life were illuminating. He remembered hearing of Lindy's 1927 flight on the radio and, as I wrote in the text, absorbed a mental toughness from his parents that saw him through the war and made him the man he became.

69   **Shortly after his inauguration:** See Okrent, *Last Call*, pp. 329–354, and pp. 373–376, for more detailed accounts. Coca-Cola, which had become vastly profitable as a manufacturer of "soft drinks" ac-tually considered producing "Coca-Cola Beer." Also, the $258,911,332 collected in newly instituted alcohol taxes represented about 9% of fed-eral revenue in 1934—an astonishing slice of the pie and enabled the president to give Americans a 20% tax cut that jolted the economy into a substantial recovery.

71   **Through it all the nation:** Richard Overy's *War and Economy in the Third Reich*, specifically Section IV, Chapter 9, "Guns or Butter," and Alan S. Milward's *War, Economy and Society 1939–1945*, Chapter 8. Both are fascinating reading for those who enjoy the background behind the fighting. For a truly detailed, exhaustively researched ex-planation of the economics of Nazi Germany I suggest *The Wages of Destruction*, by Adam Tooze. *Hitler's Eagles* by Chris McNab remains

an invaluable resource on Luftwaffe organization, training, and key personnel.

72   **Interestingly, the young British:** John Golley's *Genesis of the Jet* is a bible for all things related to Frank Whittle. See page 23 for a description of Whittle's startling 1928 thesis "Future Developments in Aircraft Design."

73   **The solution, which astonished:** Ibid., pp 32–36.

74   **Dr. Alan Arnold Griffith:** Ibid., p. 36, pp. 82–85. It is a relief, yet also discouraging, to read about Whittle's confrontations with bureaucracy and professional jealousy that often seem to be a wholly American issue.

76   **At the same time the Germans:** John Killen's *The Luftwaffe*, particularly Chapter 21, is an excellent source of detailed, hard-to-find information. See Anderson's *The Airplane*, pp. 286–289, for a summary of Germany's first jet aircraft.

78   **This outstanding achievement:** *Lords of the Sky*, pp. 175–180, describes the opening tactical moves of the Second World War.

Chapter 4: The Crucible

79   **One hour past midnight:** For a very well-written account of the German invasion of the Soviet Union from the German point of view, see *The German War*, pp. 157–199 by Nicholas Stargardt. James Holland's incomparable pair, *The Rise of Germany*, pp. 565–561 and *The Allies Strike Back*, pp. 23–41, offer insightful perspectives from the Allied side.

80   **A training slot opened:** Former test pilot Al Blackburn has a good deal of background information on George Schwartz Welch that he collected from firsthand experience, mutual acquaintances, and Welch family members. See Blackburn's *Aces Wild*, pp. 91–104. There are also several websites that, surprisingly, have solid information on Welch. The first is www.fampeople.com, and the second is a two part series from www.planesandpilotsofww2. Both are worth perusing for those wishing more information on one of the most intriguing figures of aviation history.

85   **During a single mission on October 26:** Claims that Chuck Yeager was the first ace-in-a-day in 1944 are patently untrue. German, Japanese, and Soviet pilots aside, Lieutenant Vejtasa (USN) and Lieutenant Jefferson DeBlanc (USMC) were the first Americans to gain this status in World War II. A Marine with VMF-12, DeBlanc shot down five Japanese over the Solomon Islands on January 31, 1943.

87   **January also saw:** The German Sixth Army was the largest of the Wehrmacht, and as I wrote in the text the psychological cost was on par with the physical loss of 265,000 trained soldiers. Germany's

decline had begun with the invasion of Russia, but the disaster of Stalingrad made it visual to the world.

87 **Operation Torch, the Allied invasion:** Just as Stalingrad and Guadalcanal destroyed the myths of Axis invincibility, so Operation Torch allowed a preview of what the Allies were capable of achieving. The first large-scale joint operation of the war, the invasion of North Africa was a victory in many ways. British and American planners became familiar with each other, vital logistical details were worked out and put into practice, and complex, seaborne invasions became reality. American troops in the Mediterranean Theater also gained practical combat experience that would be invaluable throughout the balance of the war.

88 **An Allied offensive:** America had been at war for less than a year and although the fighting in the Pacific was, and would remain, a bitter, vicious struggle the European/Mediterranean theater was the focus for Allied planners. The actual land area of the entire South Pacific is approximately 213,000 square miles, roughly the size of France; Japan could be contained and driven slowly back to their Home Islands, so the Pacific war would spread no further. Axis occupied Europe was at least one million square miles, and interconnected geographically so there was no telling what the Germans might attempt—the Battle of the Bulge proved that point. As long as air power guaranteed Allied freedom of ground movement in Europe while retarding Axis efforts then, like Japan, the enemy could be beaten down and crushed.

91 **"I was in the middle somewhere":** Interview with Ken Chilstrom, December 2018.

92 **The Apache was born:** This particular program had more significance than most casual readers of history realize. It was, in effect, a test bed for the P-51 version of the same aircraft. Engines, weapons, and aerodynamics were all refined through combat lessons, learned and incorporated into subsequent variants of the Mustang. See the *Allison-Engined P-51 Mustang*, pp. 4–36, for a general overview.

92 **Though there were many significant:** Anderson's *The Airplane*, pp. 236–238, has an excellent technical description of the concept while pp. 265–268 discuss the design of the P-51 itself.

100 **Ken Chilstrom and his personal:** Interview with Ken Chilstrom, December 2018. Ken was able to discuss the A-36 at length; engine settings, gun sights, and many details, including dispelling the myths about the Apache's dive brakes being wired shut. He told me this was never the case in any aircraft he flew.

101 **It was Patton himself:** Interview with Ken Chilstrom, December 2018. As I wrote, Ken still laughs over this incident. A firsthand account regarding such a prominent figure is extremely interesting and humanizes an otherwise grim situation. Ken's recollection of Bob Hope does exactly the same and permits the reader to see beyond the hype into a personal look at the man.

109   **The Germans were now:** This battle is a superb view of Germany's overall tactical mindset in microcosm. Their military was built for *blitzkrieg*—the lightning war—and they conducted their operations accordingly. This worked early on where tactical, localized, victories defeated their opponents but is not a large-scale, strategic method of conquest. The German industrial-military complex was never designed for long-term, multitheater operations, nor was any of this altered after it became necessary in 1941.

110   **Waffen SS commandos:** It is important to correct historical misconceptions whenever possible and Otto Skorzeny, by virtue of his charisma and eye for publicity, has always erroneously been given credit for this feat. Major Otto-Harald Mors, a Fallschirmjäger battalion commander, planned and commanded the raid. Georg Freiherr von Berlepsch was the first to actually free the Italian dictator. See Marco Patricelli's *Liberate il Duce!* and a work written by Mussolini's son Romano, *My Father, il Duce.*

112   **Short of a negotiated peace:** The historical context included in this chapter should make the reader aware of the German situation, and thus understand the importance of the various advanced programs. Military and political realists were keenly aware after the failure of Barbarossa, and Japan's attack on the United States that, short of technology that would level the playing field, military victory was impossible. Very soon thereafter, the more pragmatic Germans in leadership knew even the jets and rockets could not be manufactured in sufficient numbers to turn the tide—not without an atomic bomb. The Pacific theater is an interesting inversion of this situation as the Americans realized that even with overwhelming numerical superiority, virtually unlimited resources, and a truly amazing industrial capacity there was no way Japan would surrender. No way, that is, against something that could not be defended against, and so devastating that Japan faced extinction. This was precisely what led to the decision to use atomic weapons.

Chapter 5: Wonder

114   **Busemann, born in:** See Anderson, *The Airplane*, pp. 322–323. It is somewhat surprising how much was known, or at least postulated, about supersonic flight during the 1920s and 1930s. Men like Busemann and the NACA's Robert Jones had taken this to the limits of thought, yet wind tunnels, engines, and aircraft design had not achieved par with aerodynamics to the point where systematic flight

test could validate or invalidate transonic or supersonic theories. A tremendous catalyst, extensive funding, and the correct sort of pilots were needed and this occurred in 1939 with the onset of the Second World War.

116   **Termed the transonic region:** See *A History of Aerodynamics*, p. 371.

117   **This was confirmed during:** After World War II Hugh Dryden became the director of aeronautical research for the National Advisory Committee for Aeronautics then the director of the NACA until 1958. Dryden served as deputy director of NASA after it replaced the NACA, and the Dryden Flight Research Center at Edwards AFB was so named in his honor.

119   **It is interesting to speculate:** See McNab, *Hitler's Eagles*, pp. 199–200, as well as Forsyth, *X Planes: Luftwaffe Emergency Fighters*, pp. 4–23, for amplifying information. Even with early advances in jets or rockets the Reich faced a critical resource problem. Still, with a pragmatic military strategy that included securing North Africa and the Suez Canal but not invading Russia the war could well have been over before Japan's attack on Pearl Harbor.

119   **It was apparent to these men:** See Hallion's *Test Pilots*, pp. 186–188 and p. 194.

124   **It was certainly not new:** Ken Chilstrom interview, November 2017. The ability to speak to a man with firsthand knowledge removes all the guesswork or interpretation from certain historical events. As Ken related to me, and quite contrary to my previous belief, U.S. pilots like himself were not unduly concerned about German jets. They were dangerous and nearly always scored, but there too few of them to make a significant difference. Also, combat pilots will work out countertactics very fast and so it was done in Europe. Allied fighters discovered quickly that the jet could not turn well and was exceedingly vulnerable during takeoff and landing due to the long engine spool-up times.

126   **Heinkel had the obvious lead:** See Witkowski's *The Truth About the Wunderwaffe*, pp. 45–78, *Hitler's Eagles* by Chris McNab, pp. 195–204, and Forsyth's *X Planes*, pp. 24–61.

128   **By the fall there were:** The math, correlated from the *Army Air Forces Statistical Digest*, is most revealing and indisputably bears out my statements that German advanced technology did not seriously impede Allied efforts due to the timing and numbers; nor would it, short of an atomic weapon. Also, best on all sides as they were, the Germans completed only about 10 percent of the stipulated 3,000 hours of flight testing on the Me 262 and this resulted in all sorts of problems arising in operational units that were ill equipped to solve them. Turbine

blades were unreliable and gave only a few hours of service before fracturing due to the inferior metals used in construction. There were also controllability issues with elevator and aileron "jitter" at high speeds approaching the transonic region.

Chapter 6: The Brave

131   **At least 10,000 missions:** See the *Army Air Forces Statistical Digest*, December 1945.

133   **Conceived by Dr. Fritz Gosslau:** See Killen's *The Luftwaffe: A History*, pp. 241–261.

135   **Unquestionably the pinnacle:** *Lords of the Sky*, pp. 355–359, and Chorlton's *Allison-Engined P-51 Mustang*.

136   **Lockheed test pilot Ralph Virden:** For an excellent discussion of compressibility, see Peter Caygill's *Sound Barrier*, pp. 29–43.

138   **Without ball bearings:** See Charles Higham, *Trading with the Enemy*, pp. 116–130.

140   **In light of all this:** See Forsyth's *X Planes*, pp. 13–20.

141   **The forest option was:** *The Luftwaffe: A History*, pp. 231–247, by John Killen and Steven Zaloga's *Defense of the Third Reich 1941–45*, pp. 48–58.

143   **The average German lived on:** See a pair of excellent books on the subject; Richard Overy's *War and Economy in the Third Reich* and *The Wages of Destruction* by Adam Tooze.

144   **His luck changed:** Yeager and Janos, *Yeager*, pp. 32–34.

145   **"I'd rather dogfight than eat steak!":** Hoover with Shaw, *Forever Flying*, p. 56.

146   **But the Germans flying in Operation Bodenplatte:** See *Bodenplatte: The Luftwaffe's Last Hope*, pp. 227–276, by Manrho and Putz.

155   **Major George Preddy:** Preddy held the European Theater of Operations (ETO) record for kills in one day after shooting down six Bf 109s on August 6, 1944—more than two months <u>before</u> Chuck Yeager made his outrageous claim in *Yeager*, p. 72, that he was "the first ace in a day." For World War II, the Germans had six pilots who were double or triple aces in a day, including Emil Lang's astonishing tally of eighteen Soviet fighters on November 3, 1943. Other WWII aces in a day came from Japan, South Africa, Australia, Poland, and Britain, to name just a few. Before Preddy's 1944 feat, America produced five such men, all naval aviators fighting in the Pacific.

158   **Ken Chilstrom had survived:** Interview with Ken Chilstrom, December 2017.

160   **The result was a little-known:** See Robert F. Door's extremely interesting article "Project Extraversion: P-80 Shooting Stars in World

War II." Defense Media Network, April 15, 2013.

Chapter 7: World War to Cold War

167 **"Everyone knew the jet":** Interview with Ken Chilstrom, December 2017.

167 **One such operation was Lusty:** See Wolfgang Samuel's *American Raiders* and *Watson's Whizzers* for more information.

169 **"It was all stopwatches":** Interview with Ken Chilstrom, November 2017. Also refer to *Test Flying at Old Wright Field*, Chilstrom and Leary, for firsthand accounts of many pivotal test programs.

169 **Messerschmitt's P.1101 arose:** See Forsyth's *X Planes*, pp. 51–59.

172 **American Bob Jones of the NACA:** *Expanding the Envelope*, by Michael H. Gorn, is a superb source of information on flight testing and the NACA. See p. 56.

173 **"Man . . . I didn't know the 38":** Hoover with Shaw, *Forever Flying*, p. 93.

175 **American boys sporting crew cuts:** As I've said before, to understand the people in any story one must understand the context, what factors, particularly the time in which they lived, shaped them, and I utilized several excellent sources for the 1940s section of this book. *The 40s: The Story of a Decade* from *The New Yorker*; Robert Sickel's *The 1940s*, and *America in the 1940s* by Lindop and Goldstein are great places to start.

176 **At one point in 1946:** It is encouraging to both see evidence that we, and our government, can learn from past mistakes. The various veteran programs were wildly successful, beyond all expectations, in preventing a repeat of both the economic and societal issues that defined the 1920s. As the numbers bear out, an educated, trained "new" middle class arose from the cauldron of war that fueled America's ascendancy to global leadership and shaped the world we live in today.

178 **To this end George C. Marshall:** The Marshall Plan and Truman Doctrine were the pillars upon which the lines of the Cold War were drawn, and our modern world was formed. The relative success or failure of each can be endlessly debated, but what is fact is that the United States and the Soviet Union *did not* face each other in World War III.

183 **Never forgetting that the atomic bombs:** This is an excellent case study in the dynamics of military capability dictating politics, or of political reality driving military development.

Chapter 8: The Final Stage

185 **"Yes . . . I said it is":** Gorn, *Expanding the Envelope*, p. 151.

186 **A major issue was:** Hallion's *Test Pilots*, pp. 185–191, describes

this event quite well.

186   **This problem would eventually be:** Anderson, *History of Aerodynamics*, p. 412.

189   **Bell delivered its first:** Miller, *The X-Planes*, p. 24.

190   **Nineteen forty-six, then, dawned:** Interview with Ken Chilstrom, December 2017.

191   **Ken's P-80 was powered:** Ibid. A unique story that I have never seen fully recorded anywhere. Ken remembers the incident clearly and, even then, knew it was historical. He did not think much of Wright's rudeness and was personally unimpressed with the man himself.

192   **Three P-80s from:** Ibid. Another great story from a man who participated in the race itself. To hear him matter-of-factly discuss aviation legends like Robin Olds and Gus Lundquist was amazing and worth a book all by itself.

193   **"Everyone who mattered liked Robin Olds":** Ibid. For further reading on Olds, whom I also regard as a fighter pilot's fighter pilot, read Christina Olds's marvelous account of her father in *Fighter Pilot*. I also included several sections about the man in *Lords of the Sky*, pp. 346–352 and pp. 459–468.

198   **On December 9, 1946":** Miller, *The X-Planes*, p. 27.

200   **For certain, Al Boyd was no:** Too little has been written about this man. His flying skills were matched by his administrative talents and political instincts. See Chilstrom and Leary, *Test Flying at Old Wright Field*, p. 257, plus numerous other personal recollections in various accounts. He is also mentioned in Bob Hoover's *Forever Flying*, Chuck Yeager's *Yeager*, and extensively in Richard Hallion's *Test Pilots*.

202   **By late 1946:** Read Kenneth Werrell's excellent *Sabres Over MiG Alley*, pp. 3–14.

204   **In other words, the United States:** And so it began. The combination of military strength and political will on both sides created a five-decade arms race. As the post–Great War generation inherited a mess of economics, vengeance, and artificial borders so have we inherited the detritus from the Cold War.

207   **Though he was a career:** Interview with Ken Chilstrom, December 2017.

207   **In a 1989 interview:** It is worth restating that either Jack Woolams or Chalmers Goodlin could have taken the X-1 supersonic anytime in 1946 if this had been desired and contracted.

208   **"My own choice":** Fred Ascani, quoted in Yeager and Janos, *Yeager*, p. 127.

209   **"I didn't want it":** Interview with Ken Chilstrom, November, 2017.

212 **According to James Young:** *Meeting the Challenge of Supersonic Flight*, p. 46, by James Young. A very instructive read by the head of the Air Force Flight Test Center History Office.

Chapter 9: The Demon

215 **Naked, except for a pair of:** Lauren Kessler, *The Happy Bottom Riding Club*, p. 17. A wonderful book: well researched, entertaining, and informative. I intended to use it only for occasional reference and could not put the book down. Pancho and her ranch were central to the end of this story of chasing the demon because all these men, from their diverse backgrounds, war experiences, and flying careers came together here. They were, pardon the phrase, in orbit around Pancho's place; it was really their only leisure and social outlet in an otherwise dismal setting. I would surmise that Pancho knew more about what was happening in both the XP-86 and X-1 programs than Al Boyd or Stuart Symington.

220 **He was not a happy man:** According to Ken Chilstrom, this is an understatement. According to Yeager, Boyd said, "Chuck, if you let me down and do something stupid out there, I'll nail you to the cross."

224 **Bell's X-1 was finally up:** Miller's *X-Planes* has an excellent timeline of events, pp. 21–36.

227 **Like the Me 262:** This horizontal tail is basically the same used today in modern fighters and represents a synthesis of wartime aerodynamic lessons, wind tunnel data, and validations through flight test.

229 **While ferrying the plane:** Blackburn, *Aces Wild*, pp. 136–137.

230 **They all agreed that if:** Ibid., p. 136.

232 **He had a hostess friend:** Much of the personal information on George Welch comes from Al Blackburn, who was a contemporary of both Yeager and Welch. Blackburn knew the same people, hung out at the same places, and heard all the stories. I realize this is anecdotal, but it agrees with characteristics of the man himself, and the time in which he lived.

236 **According to the NAA flight test report:** North American Aviation merged with Rockwell during the early 1970s and most of the company was acquired by Boeing in 1996. I was able to procure copies of George Welch's 1947 XP-86 Flight Test Reports, courtesy of Michael Lombardi, Boeing's Corporate Historian. There are enough odd discrepancies and missing pages, along with credible secondhand accounts of October 1947, to produce very reasonable doubts about the USAF insistence that the X-1 flew supersonic first.

238 **The next few days:** See Yeager and Janos, *Yeager*, pp. 140–162, Miller, *X-Planes*, p. 28, and the National Air & Space Museum's syn-

opsis of the program. Ken Chilstrom also provided much insight on the program management aspects of that month, and the difficulties encountered then solved.

240   **By this time, B-29:** Official transcript courtesy of Dr. Alex Spencer, National Air & Space Museum. July 2017 interview.

242   **There was then, as now:** Yeager admitted this himself. "There should've been a bump on the road, something to let you know you had just punched a hole in the sound barrier": Yeager and Janos, *Yeager*, p. 165. He was quite correct; I have been supersonic thousands of times in my own career and it is a nonevent. Yeager was also correct in stating, "The real barrier wasn't in the sky but in our knowledge": He had, in one sentence, summed up the essence of the demon.

243   **"Slick Goodlin was a much better":** Interview with Ken Chilstrom, December 2017.

Appendix: Aerodynamics 101

276   For those wishing to delve deeper into this particular field, NASA is an excellent place to begin. The website is easy to navigate and has explanations at various levels that deal with all aspects of aerodynamics and propulsion. Of particular interest to those just beginning in the science is the "Dynamics of Flight" section; all the forces are openly discussed, as are the pertinent laws of motion and control.

276   **In 1738, Swiss mathematician:** Bernoulli's *Hydrodynamics*, *Sir George Cayley's Aerodynamics* by Gibbs-Smith, and Lancaster's *Aerodynamics* are all interesting reads for background purposes, though, lacking the benefits of modern wind tunnels, testing, and test pilots, the older texts are often incomplete. Still, through the reading one can gain a sense of the greatest questions and challenges from each era.

277   **In any event, the properties:** da Vinci was really the first to understand and quantify lift, as it applies to human flight. Even though the act was observed daily, and always had been, aerodynamic lift and other forces were not truly understood until the Tuscan-born polymath applied his magnificent talents to the issue. Among his writings and journals, the *Codex Trivultianus* and *Codex Atlanticus* both discuss aspects of flying and related principles. The eighteen folios of da Vinci's *Codex on the Flight of Birds* are extremely interesting, particularly Number 7, which goes into great detail on wingtip control. There is no doubt that later aviation pioneers, especially those advocating wing warping, took their ideas from these early works.

278   **This is still not flying:** Any discussion of flight, as we understand it in the modern sense, must include Dr. John Anderson's *A History of Aerodynamics*. He has named Part I of this seminal work

"The Incubation Phase" and it covers what was known, postulated, and perceived from antiquity through Cayley and the early nineteenth century. Part II is the "Infancy of Aerodynamics" and describes, in great detail, man's increasing understanding of cambered airfoils, flow, and the differences between theoretical and applied aerodynamics.

Selected Bibliography

Adams, Michael C. C. *The Best War Ever: America and World War II.* Baltimore: Johns Hopkins University Press, 1994.

Allen, Frederick Lewis. *Only Yesterday: An Informal History of the 1920s.* New York: Harper & Row, 1931.

Anderson, John D. Jr. *Computational Fluid Dynamics: The Basics with Applications.* New York: McGraw-Hill, 1995.

Anderson, John D. *A History of Aerodynamics.* Cambridge: Cambridge University Press, 1997.

Anderson, John D. Jr. *The Airplane: A History of Its Technology.* Reston: American Institute of Aeronautics and Astronautics, Inc., 2002.

Angelucci, Enzo, and Bowers, Peter. *The American Fighter.* London: Orion, 1987.

Army, Department of the. *USAAF Casualties in European, North African and Mediterranean Theaters of Operations, 1942–1946. Army Battle Casualties in World War II—Final Report.* Washington, DC: Department of the Army, GPO, 1953.

Army, Department of the. *Army Battle Casualties and Nonbattle Deaths in World War II—Final Report.* Washington DC: Department of the Army, GPO, 1953.

Army, U.S. *U.S. Army Air Forces Statistical Digest.* Washington, DC: Office of Statistical Control, 1945.

Atkinson, Rick. "The Road to D-Day." *Foreign Affairs*, July/August 2013, pp. 55–75.

Bader, Douglas. *Fight for the Sky: The Story of the Spitfire and Hurricane.* London: Cassell Military Books, 2004.

Bairstow, Leonard. *Applied Aerodynamics.* London: Longmans, Green & Co., 1920.

Baritz, Loren. *The Culture of the Twenties.* Indianapolis: Bobbs-Merrill, 1970.

Barros, Henrique Lins de. *Santos-Dumont and the Invention of the Airplane.* Rio de Janeiro: Brazilian Ministry of Science and Technology, 2006.

Bernoulli, Daniel. *Hydrodynamics* (trans. Thomas Carmody and Helmut Kobus). New York: Dover, 1968.

Bilstein, Roger E. *Flight in America: From the Wrights to the Astronauts.* Baltimore: Johns Hopkins University Press, 2001.

Bishop, Edward. *Hurricane.* Shrewsbury: Airlife Publishing, 1986.

Blackburn, Al. *Aces Wild: The Race for Mach 1.* Lanham, MD: Rowan & Littlefield, 1998.

Blackwell, Ian. *The Battle for Sicily: Stepping Stone to Victory.* Barnsley: Pen & Sword, 2008.

Blom, Philipp. *Fracture: Life & Culture in the West, 1918–1938 .* New York: Perseus, 2015.

Bob, Hans-Ekkehard. *Betrayed Ideals, Memoirs of a Luftwaffe Fighter Ace.* Cerberus Publishing Ltd, 2003.

Bowman, Martin. *P-51D vs Fw 190: Europe 1943–45.* Oxford: Osprey , 2007.

Boyne, Walter J. "Goering's Big Bungle." *Air Force Magazine,* 2008, Vol. 91.

Breuer, William B. *Operation Torch: The Allied Gamble to Invade North Africa.* New York: St. Martins Press, 1985.

Bridman, Leonard, ed. *Jane's Fighting Aircraft of World War II.* London: Studio, 1946.

Briggs, L. J., Hull, G. F., and Dryden, H. L. *Aerodynamic Characteristics of Airfoils at High Speeds.* TR 207, NACA, 1924.

Bryson, Bill. *One Summer: America 1927.* New York: Anchor Books, 2014.

Burns, Eric. *1920: The Year That Made the Decade Roar.* New York: Pegasus Books, 2015.

Busemann, Adolf, interview by Steven Bardwell. *Interview of Adolf Busemann, American Institute of Physics* (1979).

Caldwell, Donald. *JG 26 Luftwaffe Fighter Wing War Diary Volume Two: 1943–1945.* Mechanicsburg: Stackpole Books, 1998.

Caygill, Peter. *Sound Barrier: The Rocky Road to Mach 1.0+.* Barnsley: Pen & Sword, 2005.

Chanute, Octave. *Progress in Flying Machines.* Long Beach, CA: Lorenz & Herweg, 1976 (original publication in 1894).

Chilstrom, Colonel Ken. Interview by Dan Hampton. (November 5, 2017).

Chilstrom, Colonel Ken. Interview by Dan Hampton. (July 20, 2017).

Chilstrom, Colonel Ken. Interview by Dan Hampton. (November 1, 2017).

Chilstrom, Colonel Ken. Interview by Dan Hampton. (November 8, 2017).

Chilstrom, Colonel Ken. Interview by Dan Hampton. (December 10-14, 2017).

Chilstrom, Colonel Ken. Interview by Dan Hampton. (January 26, 2018).

Chilstrom, Colonel Ken. Interview by Dan Hampton. (January 31, 2018).

Chilstrom, Colonel Ken, and Leary, Penn, eds. *Test Flying at Old Wright Field.* Omaha: Westchester House, 1991.

Chorlton, Martyn. *Allison Engined P-51 Mustang.* Oxford: Osprey Publishing, 2012.

Christopher, John. *The Race for Hitler's X-Planes.* Gloucestershire: History Press, 2013.

Coombs, L. F. E. *Control in the Sky: The Evolution & History of the Aircraft Cockpit.* Barnsley, South Yorkshire: Pen & Sword Books, LtD., 2005.

Correll, John. "Daylight Precision Bombing." *Air Force Magazine,* October 2008, pp. 60–63.

Craven, Wesley F., and Cate, James L. *The Army Air Forces in World War II, Volume 2.* Chicago: Chicago University Press, 1949.

Craven, W. F., and Cate, J. L. *The Army Air Forces in World War II. Volume 6: Men and Planes.* Washington DC: Office of Air Force History,

1983.

Davis, Richard G. "German Rail Yards and Cities: U.S. bombing Policy 1944–1945." *Air Power History*, Vol. 42.

Dee, Richard. *The Man Who Discovered Flight: George Cayley and the First Airplane*. Toronto: McClelland and Stewart, 2007.

Deighton, Len. *Fighter*. London: Random House UK, 2000.

Dryden, H. L., and Briggs, L. J. *Pressure Distribution Over Airfoils at High Speeds*. TR 255, NACA, 1926.

Eisenhower, Dwight D. *Crusade in Europe*. London: William Heinemann, 1948.

Flack, Ronald. *Fundamentals of Jet Propulsion with Applications*. New York: Cambridge University Press, 2005.

Foreman, John, and Harvey, S. E. *Messerschmitt Combat Diary Me 262*. London: Crecy Publishing, 1995.Forsyth, Robert. *X Planes: Luftwaffe Emergency Fighters*. Oxford: Osprey, 2017.

Fortier, Norman "Bud." *An Ace of the Eighth: An American Fighter Pilot's Air War in Europe*. New York: Ballantine Books, 2003.

Fozard, John W., ed. *Sydney Camm and the Hurricane: Perspectives on the Master Fighter Designer and His Finest Achievement*. London: Airlife, 1991.

Funk, Arthur L. *The Politics of Torch*. Lawrence: University of Kansas Press, 1974.

Galland, Adolf. *The First and the Last: The Rise and Fall of the German Fighter Forces, 1938–1945*. New York: Ballantine Books, 1954.

Gibbs-Smith, C. H. *Sir George Cayley's Aerodynamics, 1796–1855*. London: HMSO, 1962.

Gimbel, John. "U.S. Policy and German Scientists: The Early Cold War." *Political Science Quarterly*, Volume 101, No. 3 (1986): 433–450.

Glauert, Hermann. *The Elements of Airfoil and Airscrew Theory*. Cambridge: Cambridge University Press, 1926.

Golley, John. *Genesis of the Jet: Frank Whittle and the Jet Engine*. Shrewsbury: Airlife Publishing, Ltd , 1996.

Gorn, Michael H. *Expanding the Envelope*. Lexington: University Press of Kentucky, 2001.

———. *The Universal Man: Theodore von Kármán's Life in Aeronautics*. Washington, DC: Smithsonian Institution Press, 1992.

Graves, Robert, and Hodge, Alan. *The Long Week-End: A Social History Of Great Britain 1918–1939*. London: W.W. Norton & Company, 1940.

Gregor, Neil. *Daimler-Benz in the Third Reich*. New Haven: Yale University Press, 1998.

Groom, Winston. *1942: The Year That Tried Men's Souls*. New York: Grove Press, 2005.

Gunston, Bill. *The World Encyclopaedia of Aero Engines*. Sparkford, England: Patrick Stephens Ltd., 1995.

Gyorgy, N. I. "Albert Fono: A Pioneer of Jet Propulsion." *International Astronautical Congress*, 1977.

Hague, Arnold. *The Allied Convoy System 1939–1945*. Annapolis: Naval Institute Press, 2000.

Hallion, Richard P. *Supersonic Flight*. New York: Macmillan, 1972.

—. *Test Pilots: The Frontiersmen of Flight*. Washington, DC: Smithsonian Institution Press, 1981.

Hampton, Dan. *Lords of the Sky*. New York: HarperCollins, 2014.

Hardesty, Von, and Ilya Grinberg. *Red Phoenix Rising: The Soviet Air Force in World War II*. Lawrence: University Press of Kansas, 2012.

Harvey, Henry Finder (ed.) with Giles. *The 1940s: The Story of a Decade*. New York: Modern Library, 2014.

Heinmuller, John P. V. In *Man's Fight to Fly*, by John P. V. Heinmuller, 68–85. New York: Funk and Wagnalls, 1944.

Herman, Arthur. *Freedom's Forge*. New York: Random House, 2013.

Higham, Charles. *Trading with the Enemy*. New York: Barnes & Noble, 1983.

Higham, Robin, and Williams, Carol. *Flying Combat Aircraft of USAAF-USAF (Vol. 2)*. Manhattan: Sunflower University Press, 1978.

Hoffman, Paul. *Wings of Madness: Alberto Santos-Dumont and the Invention of Flight*. New York: Hyperion, 2003.

Holland, James. *The Allies Strike Back: 1941–1943*. New York: Atlantic Monthly Press, 2017.

—. *The Battle of Britain*. New York: St. Martin's Press, 2010.

—. *The Rise of Germany 1939–1941*. London: Bantam, 2015.

Hooton, E. R. *Phoenix Triumphant* . London: Brockhampton Press, 1994.

Hoover, R. A. with Mark Shaw. *Forever Flying: Fifty Years of High-Flying Adventures from Barnstorming in Prop Planes to Dogfighting Germans to Testing Supersonic Jets.* New York: Atria, 1996.Jackson, Robert. *Fighter Pilots of World War II.* New York: Barnes & Noble, 1976.

Jacobs, Eastman. *Methods Employed in America for the Experimental Investigation of Aerodynamic Phenomena at High Speeds.* Technical paper. Washington, DC: NACA, 1936.

Johnston, A. M. "Tex." *Tex Johnston: Jet-Age Test Pilot.* Washington, DC: Smithsonian Institution Press, 1984.

Judt, Matthias, and Ciesla, Burghard. *Technology Transfer Out of Germany After 1945.* Harwood Academic Publishers, 1996.

Kármán, Theodore von. *Aerodynamics.* Ithaca: Cornell University Press, 1954.

Kármán, Theodore von. "Supersonic Aerodynamics—Principles and Applications." *Journal of the Aeronautical Sciences,* 1947: pp. 374-402.

Kármán, Theodore von (with Lee Edson). *The Wind and Beyond: Theodore von Kármán, Pioneer in Aviation and Pathfinder in Space.* Boston: Little, Brown, 1967.

Keegan, John. *Six Armies in Normandy.* Penguin Books, 1982.

—. *The Second World War.* New York: Penguin, 1989.

Keeney, D., ed. *The War against the Luftwaffe, 1943–1944.* Campbell, CA: FastPencil, Inc., 2011.

Kempel, Robert. "Mach 1 and the North American XP-86." *American Aviation Historical Society,* 2006, Vol. 51, No. 1.

Kennedy, David M. *World War II Companion.* New York: Simon and Schuster, 2007.

Kershaw, Ian. *Hitler: 1936–1945 Nemesis.* London: Penguin Books, 2000.

Kessler, Lauren, interview by Dan Hampton. *Pancho Barnes* (October 11, 2017).

Killen, John. *The Luftwaffe: A History.* Barnsley: Pen & Sword Books Ltd, 2013.

Kindleburger, J. H. "The Design of Military Aircraft." *Aeronautical Engineering Review,* 1953.

Korda, Michael. *With Wings Like Eagles.* New York: HarperCollins, 2009.

Kyvig, David E. *Daily Life in the United States 1920–1940*. Chicago: Ivan R. Dee, 2002.

Lancaster, F. W. *Aerodynamics*. London: A. Constable & Co. , 1907.

Lasby, Clarence G. *Project Paperclip: German Scientists and the Cold War*. New York: Scribner, 1975.

Lawrance, Charles L. "Air—Cooled Engine Development." *SAE Journal*, Vol. 10, No. 2 (Feb. 1922): 135–141, 144.

Lawrance, Charles L. "Modern American Aircraft Engine Development." *Aviation*, 22 (Mar. 1926): 411–415.

Levisse-Touzé, Christine. *L'Afrique du Nord dans la guerre, 1939–1945* . Paris: Albin Michel, 1998.

Linden, Dr. Bob Van der, interview by Dan Hampton. *X-1 Program* (July 21, 2017).

Lindop, Edmund. *America in the 1940s*. Minneapolis: Twenty First Century Books, 2010.

Lombardi, Michael, interview by Dan Hampton. *North American Aviation & George Welch* (October 3, 2017).

Lombardi, Michael, interview by Dan Hampton. *North American F-86 Test Reports* (October 11, 2017).

Lumsden, Alec. *British Piston Engines and Their Aircraft*. Marlborough, Wiltshire: Airlife, 2003.

Macmillan, Margaret. *Paris 1919: Six Months That Changed the World*. New York: Random House, 2001.

Manrho, John, and Putz, Ron. *Bodenplatte: The Luftwaffe's Last Hope*. Mechanicsburg: Stackpole Books, 2004.

Marrett, George J., interview by Dan Hampton. *F-86 Test Reports* (October 22, 2017).

Maurer, M. *Air Force Combat Units of World War II*. Washington, DC: Office of Air Force History, 1983.

McNab, Chris. *Hitler's Eagles: The Luftwaffe 1933–45*. Oxford: Osprey Publishing, 2012.

Miller, Jay. *The X-Planes; X-1 to X-45*. Hinckley: Midland Publishing, 2001.

Millikan, Dr. C. B. *Bachem Ba 349 Natter Interceptor Project*. Exploitation. Washington, DC: Combined Intelligence Objectives Sub-Committee, G-2 Division, 1945.

Milward, Alan S. *War, Economy and Society 1939–1945*. Los Angeles: University of California Press, 1977.

Mitchum, Samuel W. *Rommel's Desert Commanders: The Men Who Served the Desert Fox, North Africa, 1941–1942*. Westport: Greenwood Publishing, 2007.

Mitchum, Samuel W., and Stauffenberg, Friedrich Von. *The Battle of Sicily: How the Allies Lost Their Chance for Total Victory*. Mechanicsburg: Stackpole Books, 2007.

Murray, Williamson. *Strategy for Defeat: The Luftwaffe 1933–1945*. Maxwell AFB, AL: Air University Press, 1983.

Murray, Williamson. *Strategy for Defeat. The Luftwaffe 1935–1945*. Princeton: University Press of the Pacific, 2002.

National Center for Education Statistics, edited by Tom Snyder. *120 Years of American Education: A Statistical Portrait*. Washington, DC: U.S. Department of Education, 1993.

Neil, Gregor. *Daimler-Benz in the Third Reich*. New Haven: Yale University Press, 1998.

O'Connell, Dan. *Messerschmitt Me 262: The Production Log 1941–1945*. Leichestershire: Classic Publications, 2006.

Okrent, Daniel. *Last Call: The Rise and Fall of Prohibition*. New York: Scribner, 2010.

Overy, R. J. *War and Economy in the Third Reich*. Oxford: Oxford University Press, 1994.

Overy, Richard. *Battle of Britain: The Myth and the Reality*. London: Penguin , 2000.

—. *The Twilight Years: The Paradox of Britain Between the Wars*. London: Penguin, 2009.

Peczkowski, Robert. *North American P-51D Mustang*. Krakow: Stratus, 2009.

Perret, Geoffrey. *Winged Victory: The Army Air Forces in World War II*. New York: Random House, 1993.

Powers, Sheryll. *Women in Flight Research at NASA Dryden Flight Research Center from 1946 to 1995*. NASA, 1997.

Pritchard, J. Laurence. "Summary of First Cayley Memorial Lecture at the Brough Branch of the Royal Aeronautical Societ." *Flight*, November 12, 1954; 702.

Pugh, Martin. *We Danced All Night: A Social History Of Britain Between the Wars*. London: Vintage Random House, 2009.

Rickard, J. *History of War*. June 16, 2014. www.historyofwar.org/air/units/USAAF/27th_Fighter_Group.html (accessed August 10, 2017).

Rogers, Edith C. *The Reduction of Pantelleria and Adjacent Islands, 8 May–14 June 1943*. monograph, Air Force Historical Research Agency, 1947.

Roseberry, C. R. *Glenn Curtiss: Pioneer of Flight*. New York: Doubleday, 1972.

Ruwell, Dr. Mary, interview by Dan Hampton. *Theodore von Karman and Early Aviation* (November 7, 2017).

Samuel, Wolfgang. *American Raiders: The Race to Capture the Luftwaffe's Secrets*. Jackson: University Press of Mississippi, 2004.

—. *Watson's Whizzers: Operation Lusty and the Race for Nazi Aviation Technology*. Atglen: Schiffer, 2010.

Schlesinger, Arthur M. *The Coming of the New Deal: 1933–1935*. New York: Houghton Mifflin, 1958.

—. *The Politics of Upheaval: 1935–1936*. New York: Houghton Mifflin, 1960.

Schuck, Walter. *Luftwaffe Eagle*. Aachen: Helios, 2007.

Sickels, Robert. *The 1940s*. Westport: Greenwood Press, 2004.

Simpson, Christopher. *Blowback: America's Recruitment of Nazis and Its Effects on the Cold War*. New York: Weidenfeld & Nicholson, 1988.

Spencer, Dr. Alex, interview by Dan Hampton. *X-1 Program* (July 20-21, 2017).

Stack, John. "Effects of Compressibility on High Speed Flight." *Journal of Aeronautical Sciences*, 1934, pp. 40–42.

Stargardt, Nicholas. *The German War*. New York: Basic Books, 2015.

Stranges, Anthony N. "Freidrich Bergius and the Rise of the German Sythetic Fuel Industry." *Isis*, December 1984, 642–667.

Studer, Clara. *Sky Storming Yankee: The Life of Glenn Curtiss*. New York: Stackpole Sons, 1937.

Tassava, Christopher J. *The American Economy during World War II*. Backend, 2010.

Temin, Peter, and Toniolo, Gianni. *The World Economy Between the*

Wars. Oxford: Oxford University Press, 2008.

The Times. "M. Santos Dumont's Balloon ." October 21, 1901: 4.

Tokaty, G. A. *A History and Philosophy of Fluid Mechanics*. Henley-on-Thames: G.T. Foulis & Co. , 1971.

Tooze, Adam. *The Wages of Destruction: The Making and Breaking of the Nazi Economy*. New York: Penguin Books, 2006.

Townsend, Peter. *Duel of Eagles*. London: Cassell Publishers Limited, 1970.

USAF. "Transmission Transcript, First Supersonic Flight." Washington, DC: Smithsonian Air & Space Museum, October 14, 1947.

Velocci, Anthony L. Jr. "Naval Aviation: 100 Years Strong." *Aviation Week and Space Technology*, April 4, 2011, pp. 56–80.

Wagner, Ray. *American Combat Planes*, 3rd ed. New York: Doubleday, 1982.

Watkins, T. H. *The Great Depression: America in the 1930s*. New York: Back Bay Books, 1993.

Welch, George. *Flight Test Progress Report No 1*. Flight Test. Inglewood, CA: North American Aviation Engineering Department, October 1, 1947.

Welch, George. *Flight Test Progress Report No 2*. Flight Test. Inglewood, CA: North American Aviation Engineering Department, October 3, 1947.

Welch, George. *Flight Test Progress Report No 5*. Flight Test. Inglewood, CA: North American Aviation Engineering Division, October 17, 1947.

Werrell, Kenneth P. *Sabres Over MiG Alley*. Annapolis: Naval Institute Press, 2005.

Wink, Jay. *1944: FDR and the Year That Changed History*. New York: Simon & Schuster, 2015.

Wood, Derek, and Dempster, Derek. *The Narrow Margin: The Battle of Britain and the Rise of Air Power*. Washington, DC: Smithsonian Institution Press, 1990.

Yeager, Captain Charles E. *XS-1 9th Powered Flight*. Flight Test Report. Washington, DC: Headquarters, United States Air Force, October 14, 1947.

Yeager, Chuck, Cardenas, Bob, Hoover, Bob, Russell, Jack, and Young, James. *The Quest for Mach One: A First-Person Account of Breaking the*

Sound Barrier. New York: Penguin, 1997.

Yeager, Chuck, and Janos, Leo. *Yeager.* New York: Bantam, 1985.

Yenne, Bill. *The American Aircraft Factory in WWII.* Minneapolis: Zenith, 2006.

Young, James O. *Meeting the Challenge of Supersonic Foight.* Edwards AFB, CA: Air Force Flight Test Center History Office, 1997.

Zaloga, Steven. *Defense of the Third Reich 1941–1945.* Oxford: Osprey, 2012.

Zaloga, Steven J. *Sicily 1943: The Debut of Allied Joint Operations.* Oxford: Osprey, 2013.

Index

BOOKS BY DAN HAMPTON

CHASING THE DEMON

A Secret History of the Quest for the Sound Barrier, and the Band of
American Aces Who Conquered It

National Bestseller

At the end of World War II, a band of aces gathered in the Mojave Desert on a
Top Secret quest to break the sound barrier–nicknamed "The Demon" by pilots.
The true story of what happened in those skies has never been told.

THE FLIGHT

Charles Lindbergh's Daring and Immortal 1927 Transatlantic Crossing

"Outstanding. ...Riveting. ...Recommended. ...A painstaking account [that]
succeeds in placing readers in the cockpit of the Spirit of St. Louis."
—*Library Journal* (starred review)

THE HUNTER KILLERS

The Extraordinary Story of the First Wild Weasels, the Band of Maverick
Aviators Who Flew the Most Dangerous Missions of the Vietnam War

"A gripping classic. Exhaustively researched, *The Hunter Killers* puts you directly
into a Wild Weasel fighter cockpit during the Vietnam War."

—Colonel Leo Thorsness,
Wild Weasel pilot and Medal of Honor recipient

LORDS OF THE SKY

Fighter Pilots and Air Combat, from the Red Baron to the F-16

National Bestseller

"An excellent, well-researched, literate overview of 20th century warfare and
the development of the fighter plane. *Lords of the Sky* will captivate history and
aviation buffs alike."

—Stephen Coonts,
New York Times bestselling author of *Flight of the Intruder*

VIPER PILOT

A Memoir of Air Combat

New York Times Bestseller

"Offers a gripping cockpit view of modern air combat. ...Hampton is a vivid
writer and an unabashed warrior. ...An outstanding work."

—*Booklist* (starred review)

HarperCollins*Publishers*

DISCOVER GREAT AUTHORS, EXCLUSIVE OFFERS, AND MORE AT HC.COM.